Chris,
I hope y
reading as n
did writing - Bruce

DISRUPTED NETWORKS
From Physics to Climate Change

STUDIES OF NONLINEAR PHENOMENA IN LIFE SCIENCE

Editor-in-Charge: Bruce J. West

Studies of Nonlinear Phenomena in Life Science – Vol. 13

DISRUPTED NETWORKS

From Physics to Climate Change

Bruce J. West & Nicola Scafetta

Duke University, USA

World Scientific

NEW JERSEY • LONDON • SINGAPORE • BEIJING • SHANGHAI • HONG KONG • TAIPEI • CHENNAI

Published by

World Scientific Publishing Co. Pte. Ltd.

5 Toh Tuck Link, Singapore 596224

USA office: 27 Warren Street, Suite 401-402, Hackensack, NJ 07601

UK office: 57 Shelton Street, Covent Garden, London WC2H 9HE

British Library Cataloguing-in-Publication Data
A catalogue record for this book is available from the British Library.

Studies of Nonlinear Phenomena in Life Science — Vol. 13
DISRUPTED NETWORKS
From Physics to Climate Change

Copyright © 2010 by World Scientific Publishing Co. Pte. Ltd.

ISBN-13 978-981-4304-30-6
ISBN-10 981-4304-30-1

Printed in Singapore by Mainland Press Pte Ltd

Preface

This is a book about complexity, complex networks and how their smooth dynamics is often disrupted. But before we can proceed it would appear that we should answer the question: "What is complexity?" Over the past two decades both professional scientists and lay people alike have wondered about the scientific meaning of this simple yet elusive word. In the fourth century St. Augustine asked himself a related odd question: "What is time?" His answer was astonishingly interesting:

> What, then, is time? If no one asks me, I know what it is. If I wish to explain it to him who asks me, I do not know.

A similar answer can be reformulated about the concept of complexity. OK, let us see how it sounds: "What, then, is complexity? If no one asks me, I know what it is. If I wish to explain it to him who asks me, I do not know." But science does not wait for definitions, science continues forward in its investigations of a phenomenon with or without clear understanding, confident that such understanding will eventually come.

As with the concept of time, the concept of complexity cannot be explained or defined in a simple way. It is easier to appeal to intuition, that is, to that mysterious faculty that allows humans to visualize the meaning of a difficult concept without confining it to a definition. Thus, we decided in writing this book that an interesting way to stimulate the intuition of our readers about the nature of complexity and complex networks is to give a concise overview of how the scientific, technological and sociological facts that emerged since the end of the twentieth century have engendered the need to address a significant portion of what is viewed as science from a new perspective. This new perspective does not rely on any particular discipline for its articulation and is known as the science of complexity. Our interest, in particular, is in its special form as the science of complex networks.

We found the approach of viewing the changes in sciences from above to be quite interesting and stimulating and hope you do as well. We realized in writing that it might be possible to describe what complexity is, and what complex networks are, using a language that can certainly attract the attention of a wide range of non-professionals. Adopting an appropriate didactic approach to explain complex networks is not, however, just an attempt to reach people who are not familiar with the language of science. Indeed, finding a way to communicate effectively with most educated people inside and outside the scientific community is a necessity given society's reliance on science and technology. Such a language might also entice professional scientists from a multitude of different fields, such as neurophysiologists, biologists, sociologists, meteorologists, chemists and, of course, physicists to work together on problems of importance to society.

The image of the renaissance person, the universal scientist such a Leonardo da Vinci who was able to master all the known science of his day, has faded from the possible. This loss over the last century is due to the development of a new kind of science that requires ever increasing specialization. This need to specialize birthed scientists so deeply compartmentalized that they isolated themselves from each other, and by doing so they isolated their fields of research as well. Mathematics, physics, biology, geology, astronomy, sociology, economics, medicine and even their sub-disciplines are now believed to be intrinsically separate domains of knowledge. As for a way to communicate, isolation yielded to differentiation, and what were just small dialectical variants at the beginning, in a few scientific generations evolved into specialized languages that have made each field of research foreign to all but the expert.

However some scientists realized that specialization in just one field of research was not always beneficial, and could, in fact, be a fatal limitation on knowledge. Understanding complex networks requires knowledge that does not reside within a single discipline and, because of the extreme richness of the current level of science, requires collaboration across disciplines. In his book *Cybernetics*, the mathematician Norbert Wiener observes:

> ...a proper exploration of these blank spaces on the map of science could only be made by a team of scientists, each a specialist in his own field but each possessing a thoroughly sound and trained acquaintance with the fields of his neighbors; all in the habit of working together, of knowing one another's intellectual customs,

and of recognizing the significance of a colleague's new suggestion before it has taken on a full formal expression.

Multidisciplinary collaboration requires a common language: a kind of *lingua franca*. The only lingua franca that is available is the non-technical language of ordinary speech. Thus, one problem scientists face is how to use a non-specialized language to communicate and introduce the progress made by scientists in different fields on the understanding of complexity and complex networks. It is the notion, role and function of scientists that is continuing to change in a rapid manner; that being not only investigator, but communicator as well. The necessity for better communication within science and between science and society is forcing new ways to illustrate scientific progress.

Thus, we organized our book in such a way as to bring the reader along on a wonderful trip through the emergence, growth and expansion of modern science, the Science of Complex Networks. We guide you through several examples from different areas of complexity, rather than providing a chronological review of what has been done. Because of this diversity the exemplars may appear disconnected. But with each stop on the junket, we believe that you will see more and more clearly how the apparently disconnected stories and findings are intrinsically linked by a novel underlying scientific methodology.

You will be able to recognize that a systematic investigation of complex networks has emerged as a new kind of science; based on a new scientific methodology. What will become evident is an irreducible difference between the dialectical two-tiered structure of the traditional scientific methodology (experiment and theory) and the new methodology, which naturally emerges in the study of complex networks that entails three tiers: experiment, computational simulation and theory. This new methodology can be transcribed into data, information and knowledge.

The historically two-tiered science is how the scientific method is usually presented and understood. The data resulting from observation/experiment suggest a theoretical model that not only explains the data but also enables prediction. Such predictions are tested by doing new experiments and/or making new observations that lead to improved theoretical models. This dialectic process between successive improvements on experiments and theory leads to an iteratively progressive understanding of a given phenomenon. Until recently all major fields of science developed by following this two-tiered

science methodology. The two-tiered science allowed us to discover fundamental laws of physics and show how the vanishingly small and the astro-
nomically large are part of the same unity. However, what is most interesting
is the realization that this methodology, which facilitated the development
of science from the time of Newton, works only if the theoretical predictions
can be directly tested against the observations. This obvious constraint is
relatively easy to satisfy only if the network under study is simple. Thus, the
historical methodology necessitates isolating or disentangling elements from
the whole and it is the ability to disentangle that makes the phenomenon
simple.

However, typical complex networks, such as those found in biology, geophysics, sociology and economics, cannot be disentangled into elementary
components to be studied separately. What makes a complex network complex is the fact that it is an entangled organism. It is the structure of these
networks that characterize them, altering the topology of complex networks
changes them in a disruptive way. Cutting the heart out of a dog to better
study how it works may not be satisfactory because, after all, the result is
a dysfunctional organ and a dead dog! Even if the heart is kept alive artificially the dog is still dead. Thus, studying complex networks requires a
middle ground to fill the gap between, say the theoretical understanding of
fundamental physics and the often poorly resolved experimental observations
of biology. This filling-in is done with complex calculations and computer
simulations that have been made possible by the increasing availability of
computers and the enhanced complexity of computer algorithms over the
last few decades. It is this computational complexity that constitutes the
third level of the new scientific methodology that is required for studying
complex networks.

Three-tiered science is a new methodology because the dialectic form
of the traditional two-tiered scientific method is disrupted by the alternatives presented by this additional level of investigation. The viability of the
third tier has been continually tested by scientists over the last half century.
Theory and experiment can no longer be directly compared in many studies because the phenomena are too complex, and computational complexity
does not establish the uniqueness of such a comparison. Different alternative
models can be opportunely tuned or adjusted to make them fit the data, but
it is not possible to start from first principles and determine which model
is the better representation of reality. Significant analyses of the data is
required to determine which model is preferable and sometimes these data

processing efforts are frustrated by the low quality of the data. Consequently, competitive theories (models) are not really tested against each other with the kind of certainty that has historically characterized such comparisons in theoretical physics.

We decided to illustrate the procedures involved in analyzing existing complex networks by discussing the debate on climate change. Few networked phenomena are more complex than the Earth's climate. Even fewer networked phenomena generate issues that are more intriguing than that of climate change and global warming where the tree-tired science of data, computer simulations and theoretical knowledge is so well exposed with all its benefit and difficulties. This debate is not an arcane academic exercise but is an important issue of general interest to most industrialized societies that are concerned about the future of our planet. The example of climate change is also a useful illustration of the influence of society on science, in terms of what research is supported, as well as, the influence of science on society in terms of what phenomena are thought to be important. Consequently, the climate change example, which is extremely important in itself, herein becomes a paradigm of a disrupted complex network exposing the strengths and weaknesses of this nascent science.

This book highlights a number of features concerning how science really works and not necessarily how we would like it to work. Complex networks, complexity, commonality, interdisciplinarity, transdisciplinarity, *etc.*, are all important new ways of looking at the world. Network science is presented as a new kind of epistemology, or way of knowing the world; not just the world of physical science, but the world of biological, economic, social, and life sciences as well. We hope that our discussion will assist lay readers in understanding how the young field of complex networks is evolving and will provide the professionals in many different areas of research a perspective by which to appreciate the interconnectedness among all disciplines. This book results from the efforts of two physicists with multiple interests and is not a substitute for a textbook, but it may supplement such texts with a pleasurable and educational read. Hopefully it will bridge the gaps among a variety of disciplines that the three-tiered science of our times entails.

Bruce J. West
Nicola Scafetta
Physics Department
Duke University, Durham NC

Contents

Chapter 1

Why a Science of Networks?

The idea that a country's government could and should financially support science for the good of society was born during the second world war. The seed of the idea was planted by President Roosevelt, in a letter to Vannevar Bush, an MIT professor who was doing his part for the war effort by serving as the Director of the Office of Scientific Research and Development (1941-1947). The president, at the height of the war, sent a letter to V. Bush, asking him how the substantial research that was being accomplished during the war could be carried over to peace time and applied to the civilian sector. V. Bush's response was one of the most influential unpublished reports in history, *Science-The Endless Frontier* [20]. In this now legendary document V. Bush argued that the United States needed to retain the scientific advantage achieved during the war years and laid out the reasons for building a civilian-controlled organization for fundamental research with close liaison with the Army and Navy to support national needs and with the ability to initiate and carry to fruition military research.

V. Bush emphasized that historically scientists have been most successful in achieving breakthroughs when they work in an atmosphere relatively free from the adverse pressure of convention, prejudice, or commercial necessity. This freedom from hierarchical structure stands in sharp contrast to military tradition. He believed that it was possible to retain an alternate organizational structure, outside the more traditional military, but working in close collaboration with it. Such an organization would foster and nurture science and the application of science to new technologies, through engineering. In V. Bush's words:

> ... such an agency should be ... devoted to the support of scientific research. Industry learned many years ago that basic research cannot often be fruitfully conducted as an adjunct to or a subdivision of an operating agency or department. Operating agencies have immediate operating goals and are under constant pressure to produce in a tangible way, for that is the test of their value. None of these conditions is favorable to basic research. Research is the exploration of the unknown and is necessarily speculative. It is inhibited by conventional approaches, traditions and standards. It cannot be satisfactorily conducted in an atmosphere where it is gauged and tested by operating or production standards. Basic scientific research should not, therefore, be placed under an operating agency whose paramount concern is anything other than research.

His vision was manifest through the founding of the *Office of Naval Research* in 1946, the *Army Research Office* in 1951 (as the *Office of Ordinance Research*), the *Air Force Office of Scientific Research* in 1950 (as the *Air Research and Development Command*) and the *National Science Foundation* in 1950; albeit, none of these organizations followed all his suggestions regarding the management of scientific personnel and the support of science. The voice of Vannevar Bush concerning the incompatibility of fundamental research and mission agencies was prophetic. The dire consequences of that incompatibility were held off for over half a century, however, by a set of safeguards put into place in order to insulate basic research (called 6.1 research in the military) from the pressures of applied research (called 6.2 research) and the fielding of technology (called 6.3 and higher research). However, Bush's cautionary voice is now being echoed in a report [27] authored by members of the National Research Council for the Department of Defense (DoD). The findings of the report of most relevance to the present discussion are:

> ... A recent trend in basic research emphasis within the DoD has led to a reduced effort in unfettered exploration, which historically has been a critical enabler of the most important breakthroughs in military capabilities. ... Generated by important near-term DoD needs and by limitations in available resources, there is significant pressure to focus DoD basic research more narrowly in support of more specific needs.

These cautionary observations are even more germane as we enter a time in world history where fundamental research for understanding is not just in the traditional disciplines of physics, biology, sociology, etc., but is spread out to the more complex phenomena that arise in world-wide transportation, international communications, global business, and the world's energy. These trans-disciplinary phenomena and others like them have been gathered together under the rubric of "networks", and their study under the label of "network science", even though the word network remains vague and whether or not a science can be constructed remains to be seen.

If a network science is ever to be formulated, scientists must learn how to couple successive scales in physical networks, from the atomic, to the molecular, to the mesoscopic, to the macroscopic, which they cannot do presently. This understanding of across-scale coupling may then be applied to the more general network-of-networks multiscale phenomena found in the biological, informational, physiological and social sciences; the DoD included. In the network-of-networks, every scale is linked to every other scale, either directly or indirectly, and a science capable of predicting the influence of changes across these multiple scales on network operations is not only desirable it is mandatory. In order to carry out this ambitious plan it is necessary to have a map of what we now understand about complex networks and where the unknown and unexplored regions are. The formulators of this new science must be aware of the desert areas where Generalized Systems Theory lies abandoned, the depth of the abyss into which Cybernetics crashed and the locations of the nomadic regions where the scientifically disenfranchised continue to walk in limit cycles. In short, we need to understand the scientific barriers that must be surmounted in order to achieve a Science of Networks.

V. Bush's warnings also have modern application in a domain he had not anticipated. It is not only the short-term vision of mission agencies that are influencing the results of science; applications are systematically identified and balanced against the results of more long-term fundamental research. World-wide politics in a variety of forms including the United Nations is influencing, if not guiding, the 'scientific interpretation' of what some consider to be the most social challenging research problem of our time — *climate change*. We use global warming as an exemplar of both the science of complex networks as well as the manner in which important questions influence and are influenced by social networks. Moreover these considerations concern the more general question of whether a science of networks is possible and if it is, what form it might take.

1.1   The science of data, information and knowledge

Some DoD scientists spend their time thinking about where science is going and whether it will benefit the military and the country when it gets there. Others investigate how new scientific understanding can result in ways to help the soldier better survive while simultaneously making him/her better able to carry out a variety of missions. These latter considerations are in the best tradition of Leonardo da Vinci (1452-1519), the legendary Florentine, who made his living designing armaments for Italian City States, and staging elaborate parties for his benefactors, as well as by painting. The former group of scientists have a job that is not as well focused as the latter and requires thinking long term to identify emerging areas of science that can positively contribute to medicine, the development of new materials, enhance communications, and augment the design of computer software and hardware. In short, this group is dedicated to finding ways to facilitate the development of the science being pursued in industry, in the academy and in government laboratories to enable the future defense and well being of the country. Science in response to the immediate needs of the soldier and science focused on the long-term needs of humanity are the extremes on a continuum of complex phenomena that scientists are attempting to understand, both in and out of the government.

Exemplars of the two kinds of science are found throughout Leonardo's Notebooks and Figure 1.1 gives a dramatic contrast of evolutionary and revolutionary science. The giant cross bow in Figure 1.1a was designed to be so large that its bolt could breach the wall of a castle; note the man in the middle foreground providing a reference scale. This kind of evolutionary science has a ripple effect on technology, leading to incremental advances on what is already known. The impact of these technical contributions is cumulative, often taking decades for their realization in marketable technology. By way of contrast Leonardo's design of a helicopter in Figure 1.1b, based on a child's toy, typifies revolutionary thinking that required four hundred years for the rest of science to catch up and build a prototype. This revolutionary science often has a tsunami effect resulting in the development of totally new strategic and tactical thinking that disrupts the status quo.

Today's most common theater of war is the downtown district of any city that shields combatants in civilian clothes. The technology of urban warfare

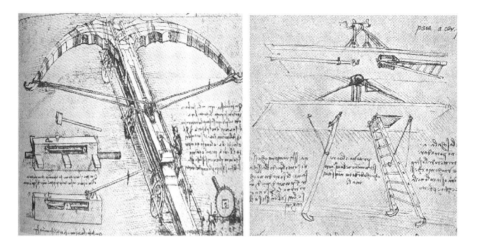

Figure 1.1: Two drawings copied from Leonardo da Vinci's *Notebooks*: (a) giant cross bow; (b) helicopter, based on a child's toy. (Copied from [134] with permission).

is not that of traditional tanks and bombs. In modern warfare it is not politically, economically nor humanely acceptable to indiscriminately bomb a city in an enemy's country, as it was in the Second World War. The warfighter must determine in which houses the true enemy resides and destroy only those buildings. Command must deploy soldiers in an urban environment who are trained to selectively capture or kill only certain members of the indigenous population while being respectful toward all others. It is no longer the case that one's enemy is on one side of a line and one's friends are on the other: patterns that recall the simple Euclidean shapes of lines, arcs, triangles, squares and circles. Today, the enemy might be both far away and in the house next door, the two mix together like the fractal fingering of a drop of ink in water (see Figure 1.2). This is part of the new complexity of warfare that is being addressed in the twenty-first century and requires a new kind of situational awareness on the part of the civilian population as well as the military.

Non-Euclidean warfare is only one instance of the fundamental peculiarities of today's world. Many believe that this complexity arises just because of a *clash of cultures*. But different cultures came into conflict at many times in world history. Therefore, although a clash of cultures might be a contribu-

Figure 1.2: The complex geometries generated by the spreading of a drop of ink in the water. These geometries might be qualitatively similar to the real disposition of today's urban warfare zones with their complex interconnections and paths. (from *http://www.flickr.com*)

tory cause to many of today's conflicts it does not explain the *form* in which these conflicts occur. Today's warfare, rather than being anomalous, seems to reflect the intricateness and complexity of the networks that permeate modern society.

Modern society, and by that we mean mainly western society, is more interconnected than societies have been in the past [28]. A western city could not function without garbage collection, interconnecting sewers and waste treatment, electricity from the power grid, transportation networks, food distribution, health care networks, and it would have a much different form without networks of education, banking, telephone service and the Internet. These activities are supported by physical and social networks within the city and their forms have been evolving for millennia. Part of that evolution was the development of their inter-operability such that these networks are all interconnected and in one way or another they connect to national and eventually to global networks. This network-of-networks is the engineered webbing of humanity, but there are comparable structures in the biosphere and ecosphere involving plant and animal network-of-networks of tremendous variety. Consequently, we believe that clues to understanding human-based complex adaptive networks may be found in naturally occurring complex networks.

It is not only our external world that is cluttered with networks, but our internal world as well. The neuronal network carrying the brain's signals to the body's physiological networks is even more complex than the modern city or a typical ecological network, if there is such a thing as a typical complex adaptive network (see Figure 1.3). The biological/ecological-networks are certainly as difficult to understand as the physical/social-networks. It is premature to assign preeminence to any one complex adaptive network or collection of such networks from any particular discipline before we understand the patterns formed by the multiple interactions within and among them.

What we discuss in this book is the evidence that there are common features for the various kinds of networks and this commonality might be exploited for the development of a science of complex adaptive networks. If such a science exists it would be very different from traditional disciplines, and we address that point subsequently. Thus, although there has been an avalanche of research into a variety of networks over the past decade, the recognition that we need an overarching science of networks is of even more recent vintage. Network science suggests a new way to discuss data and

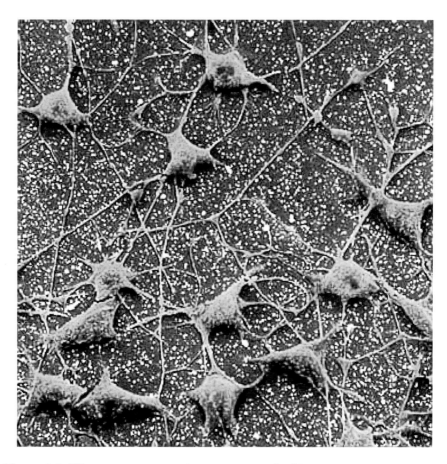

Figure 1.3: The nervous system is a vast network of interconnecting neurones which transmit information in the form of electrical signals. Our brain has around 100 billion neurones, and each communicates with thousands of others (Photo from *Science Photo Library*).

how data is fused together, such as within the human brain. *Data*, as the nineteenth century empiricists discussed, is the raw sensory material that processing transforms into *information* and, finally, the interpretation of the information produces our *knowledge* about specific phenomena.

In the present context the social domain is only one of the many realms dominated by networks, however we find that much of our present understanding of networks is based on research conducted on the social latticework. We briefly review some of the misconceptions of the nineteenth century and show how uncertainty has been quantified and made systematic in today's models. The data extracted from social nets is processed to determine patterns in terms of probabilities, which form the basis of social information. The probability functions (information) are used to predict the possible futures of the network, whose regularity forms the knowledge upon which decisions are made. This book investigates some of the gaps that exist in our understanding of the interrelation among data, information and knowledge; gaps which must be closed in order to achieve a science of networks.

If it were only a matter of evolutionary changes scientific progress would still be difficult but would be doable and would advance smoothly without any surprising outcomes. In fact, it might even be possible to predict the new technologies that most certainly will emerge from new scientific understanding. However, some of the barriers to understanding require more than new technologies; they also require new ways of thinking, such as entailed by Leonardo's disruptive designs in the distant past.

An example of a disruptive technology is the telephone invented in 1856 by the Italian Antonio Meucci (1808-1889), as the US Congress recognized in 2002 (see Figure 1.4). During the twentieth century the telephone decreased the time interval for long-range communication from weeks or months to seconds or minutes and was responsible for the development of the largest information network since the ancient Romans developed the road or more properly the "inter-state" highway.

It could be argued that the electronic computer, from the high-performance super computing devices to the common personal-computers, was an even more disruptive technology than the telephone, but the computer has done much more than accelerate communication, particularly in the sciences. The computer enormously accelerated numerical computation and, therefore, allowed the study of phenomena too complex to be solved with pencil and paper. The numerical study of complex networks has paved the way to the most important methodological scientific revolution since Galileo and New-

Meucci's telephone,
Said to have been made in the year 1857,
As reported in *Scientific American*, 1884.

Figure 1.4: Antonio Meucci, the inventor of telephone. On 11th June 2002 the U.S. House of Representatives stated that "the life and achievements of Antonio Meucci should be recognized, and his work on the invention of the telephone should be acknowledged" (H. Res. 269). Antonio Meucci was unable to raise sufficient funds to pay his patent application so that in 1876 Alexander Graham Bell was legally allowed to patent the telephone with his name that generated a controversy that birthed the single most litigated invention in U.S. history.

ton and, as such, is yielding an ongoing revolutionary strategy for using data, extracting information and developing knowledge.

As with all revolutionary events in the history of science the computer is accompanied by technical problems, theoretical difficulties and controversies whose solutions likely require novel ways of thinking, but we postpone advancing that discussion until later. The disruptive or revolutionary science that spawned the computer began with the development of information theory, cybernetics and the theory of communication. Like a tsunami the original changes due to information science were small and readily assimilated, but as time passed the science-induced technological changes became more substantial, perhaps even socially maladaptive, and eventually crashed down on humanity.

Disruptive science is implicit in the apparently benign distinctions among data, information and knowledge. If we could provide a simple description of the differences between these three entities there would be no need for this book. In fact, the confusion over the difference between data and information, as well as between information and knowledge, may well lead to the destabilization of the conceptual underpinnings of western society. Examples resulting from this destabilizing influences is the emergence of global terrorism, the spreading of the global recession of 2007-2009, and the growth in the concern over global warming to name a few. Of these examples we elect to discuss global warming in detail to clarify the implications of network science.

Global warming refers to the nearly one degree Centigrade increase in the average global temperature over the past 100 years. This phenomenon is distinct from the causal attribution of this warming to the human release of pollutants into the atmosphere. Thus, the change in the average global temperature is a physical fact, but the cause of this change is presently subject to interpretation and consequently to controversy. This example was chosen because of its potential impact on society and the passion with which the various participants hold their views. One of the things we attempt in this book is to show why the accepted causal interpretation of this phenomenon is the result of confusion and exploitation over what constitutes data, what is meant by information and most importantly, what actually constitutes knowledge. These same ideas can be brought to bear on the social phenomenon of terrorism networks and what can be learned about them from an understanding of complex adaptive networks.

1.2 The face of science

It is not fully appreciated today that the term science was invented by its practitioners in the nineteenth century to distinguish themselves and what they did from natural philosophy. Many who today we consider the fathers (and mothers) of science were, in their day, natural philosophers, and that, for better or worse, is no longer true. There was a transition from the philosophical, some might say theoretical, way of asking questions of nature to the scientific formulation of performing experiments. This is where the mythical notion of the isolated individual working alone in the laboratory was created. But science is a social activity in which the ideas of individuals are shared, critiqued, developed, refined and shared again. Science is a very personal human activity in which the thinking and understanding of the individual is made public and subjected to analysis and sometimes to ridicule. Working scientists understand the creative process as it relates to science and very often as it relates to the arts and humanities, as well. However, public exposure of one's scientific work, without the de-facto backing of the scientific community, is another matter altogether and is part of the reason why scientists are reluctant to consciously embark on the development of a new science.

Consequently, work done as a scientist requires that assertions and speculations be consciously and explicitly separated from reliable conclusions drawn from theory used to interpret the results of experiment. The purpose here is to explore one of the possible directions science may take in the near future. Of course, in such an enterprise it is necessary to speculate, make assertions and generally run the risk of looking foolish.

The arguments scientists present to convince other scientists are much different from the arguments given here, where the intended audience is the informed non-scientist that must, nonetheless, make what are ultimately scientific decisions. Society requires that such decisions be made by politicians and judges, who have no special training in science. So here we begin with the romantic notion that the face of science is a composite of the faces of the scientists whose work we have all heard about. We do this so when you evaluate our remarks it can be done with a knowledge of our perspective regarding the value of science and its place in society.

Many know the names Jonas Salk (1914-1995), the American scientist who developed the killed vaccine that eradicated polio; Louis Pasteur (1822-1895), the French physician who developed the vaccine against rabies and taught us to purify our milk by slowly raising its temperature; Albert Einstein (1879-1955), the Austrian physicist who revolutionized how we think about space and time; and Thomas Edison (1847-1931), the self-taught American who invented hundreds of new devices including the electric light and the phonograph. These individuals, along with many others, collectively constitute the public face of science. Some individual faces, like Einstein's, we know for what might be considered esoteric reasons and others, like Edison's, because they literally changed what our lives would have been like without him, in clearly defined ways.

The purist might object to putting Edison and Einstein into the same category, since the former was a consummate inventor and the latter was an unsurpassable theorist. But these men shared a common spirit; a spirit that breathed life into the social structure of the twentieth century. Einstein is often perceived as living almost completely within his own head, periodically letting others peer inside to see what he was thinking by publishing a paper.

Figure 1.5: The faces of science: Jonas Salk (top-left), Albert Einstein (top-right), Louis Pasteur (bottom-left), and Thomas Edison (bottom-right).

Leaving much of the legend aside, we know that young Albert was not a remarkable student and could not get an academic job in science after he obtained his Ph.D. in 1900, except as a part-time lecturer. Consequently, Einstein accepted the position of clerk in the Swiss Patent Office in 1901, where he evaluated the technical merit of patents; often explaining that a perpetual motion machine violates the second law of thermodynamics to a would-be inventor. Some believe that it was this relatively intellectually stress-free existence for the eight years he was at the Patent Office that enabled him to indulge his scientific imagination and work in his spare time on revolutionizing theoretical physics. Einstein published three classic papers in the Annals der Physik in 1905, after being a clerk for four years, any one of which would have been the crowning achievement of a lesser scientist's life work. The 1905 paper on the photoelectric effect earned him the Noble Prize in 1923; the 1905 paper on special relativity made him a scientific legend and folk hero and the 1905 paper on diffusion was to motivate Perrin to do experiments for which he (Perrin) would win the Noble prize in 1926. But Einstein's *academic* career did not officially begin until 1909 when he was offered a *junior* Professorship at the University of Zurich.

Edison was a different sort of man altogether. It is estimated that he had no more than three months formal education before his mother began home-schooling; yet he was to file 1093 United States patents, the most patents issued to any single individual. He also set up the first modern research laboratory at Menlo Park, New Jersey in 1876. At a time when most individuals were finishing high school and either making plans for college or getting a job, he, because of his nearly complete deafness and aversion to mathematics and theory[1], was conducting his own physics and chemistry experiments. These experiments were done using money Edison obtained through his own resourcefulness, such as writing, publishing and selling his own newspaper on a local commuter train. His investigations spanned the full spectrum of technology; he invented the first electric light bulb, the first phonograph and the first talking pictures. His ideas strongly influenced society, for example, he set up the first electrical power distribution company. Edison was as imposing a folk hero in technology as Einstein was to become in the pure sciences.

---

[1] At a young age he began reading Newton's Principia but set it aside because of its turgid and arcane style of writing. Because of this experience he decided that theory was a waste of time, obscuring rather than clarifying one's understanding of physical phenomena.

But, these are two of the better-known public faces of science, and herein we do not distinguish between basic science and technology. We intentionally leave vague the boundary between the two in order to avoid a discussion that would obscure our more general purpose. It will be to our advantage to retain a view of science that encompasses knowledge as sought by both Edison and Einstein, but without being overly concerned with the tools employed. Our purpose is to reveal the contributions of a number of scientists that have made remarkable, if less well known, discoveries that have determined how we view the modern world, particularly our understanding of network-of-networks. So, let us briefly introduce the star of this little drama and what it was that he contributed to the human store of knowledge.

The Marquis Vilfredo Frederico Damaso Pareto (1848-1923) was one of those scientists whose investigations have markedly influenced our understanding of modern society and yet, with the exception of a relatively small number of students of economics and sociology, his name is unknown. Most people know little or nothing about him or what he accomplished. His scientific work, like those of his contemporaries, Edison and Einstein, still influences us today, but in ways that have remained largely in shadow. His interest was not the mathematically intense modeling of the physical world, that motivated Einstein; nor was it the technologically new and economically viable that motivated Edison; his interest was to use the mathematical and quantitative methods of physics and engineering to understand everyday social phenomena such as why some people make more money than others. He was neither a scientist nor an inventor; he was an engineer.

Money and sex are topics that people never tire of discussing, debating and arguing about. At a certain stage of life men love to advertise their level of sexual activity to other men, but as they grow older, marry and have children, these discussions are replaced with equally passionate arguments about money and how unfairly money is distributed within society. Those that *have not* scheme how to get it and those that *have* endlessly plan how to hold on to it. The fundamental non-egalitarian imbalance in society between the rich and poor has always been present, but the arguments, until very recently, explaining the schism, have been literary, philosophical and moral; not scientific. Political arguments set one segment of society against the other, pointing out the unfairness of the status quo and claiming that the imbalance can be reduced to zero, if only this or that social theory is adopted. The argument for this Utopia, in which all people earn the same or nearly the same amount of money, is based on philosophy and not on science.

Figure 1.6: Marquis Vilfredo Frederico Damaso Pareto (1848-1923), engineer, sociologist, economist and philosopher. He made several important contributions especially in the analysis of individuals' choices and in the study of income distribution, which was discovered to follow for high income an inverse power law: $P(w) = 1/w^{\alpha}$, where w is the level of income and $P(w)$ is the relative number of people having that income. He introduced the concept of Pareto efficiency and initiated development of the field of microeconomics.

The non-egalitarian imbalance in the distribution of wealth, which seems to fly in the face of the democratic notion of equality, appears historically to be characteristic of stable societies. The scientific position regarding the necessity of this imbalance dates back to Pareto, who was the first person to use data to quantify the imbalance in a universal way. But let us postpone learning more about Pareto until we have looked a little more closely into the history of science and examined how physics has influenced our understanding of complex networks in the social and life sciences. Physics is considered by many to be the paradigm of science, so understanding how its techniques have been previously applied may guide our strategy for developing a science of networks; both to those applications that worked and those that did not. In fact it was the failures of the physics paradigm to explain complex systems in the social and life sciences that provided the first hints of what is required for a science of networks.

1.2.1 Qualitative and quantitative

One of the pioneers of quantum mechanics, Ernest Rutherford, once observed that physical theories are, and should be, quantitative in the following way:

> All science is either physics or stamp collecting. Qualitative is nothing but poor quantitative.

This view of the foundation of science is shared by a majority of scientists and from our experience it reflects the feelings of the scientific community as a whole. The mathematician Rene Thom [173] elaborated on this perspective by pointing out that at the end of the seventeenth century there were two main groups in science, those that followed the dictates of Descartes and those that accepted and practiced the physics of Newton:

> Descartes, with his vortices, his hooked atoms, and the like, explained everything and calculated nothing. Newton with the inverse power law of gravitation, calculated everything and explained nothing. History has endorsed Newton and relegated the Cartesian construction to the domain of curious speculation.

Here we see the emphasis Thom placed on calculation in the physical sciences, as distinct from explanation in the sense of creating new knowledge. A deeper probe into the numbers calculated is made by Bochner [15]:

> The demand of quantitativeness in physics seems to mean that every specific distinction, characterization, or determination of state of a physical object and the transmission of specific knowledge and information, must ultimately be expressible in terms of real numbers, either single numbers or groupings of numbers, whether such numbers be given "intensively" through the medium of formulae or "extensively" through the medium of tabulation, graphs or charts.

From these considerations one could conclude that if an explanation is not quantitative, it is not scientific. This conclusion formed the visceral belief that molded the science of the twentieth century, in particular, most of those theories emerging in the disciplines relating to the social and life sciences. Much has been written of both the successes and failures of applying the above dictum outside the physical sciences, but here is not the place to review that vast literature. Instead let us concentrate on the successes of applying the non-traditional perspective that qualitative can be as important as quantitative in understanding complex phenomena, particularly complex adaptive networks. The most venerable proponent of this view in the last century was D'Arcy Thompson [174], whose work motivated the development of catastrophe theory by Thom. The interest of Thompson and Thom in biological morphogenesis stimulated a new way of thinking about change - not the smooth, continuous quantitative change familiar in many physical phenomena, but the abrupt, discontinuous, qualitative change familiar from the experience of "getting a joke," "having an insight" and the "bursting of a bubble."

Catastrophe theory has its foundation in topology and is therefore qualitative rather than quantitative, which is to say the theory deals with the *forms* of things and not with their *magnitudes*. For example, in topology all spheres are equivalent regardless of their radii, consequently the earth and an orange are topologically indistinguishable. Furthermore, all shapes that can be obtained by smoothly deforming a sphere are also topologically equivalent. Thus, a sphere and a bowl are the same, but a cup with a handle is different since the handle has a hole in it. However, a cup and a bagel, or any other one-holed shape are topologically equivalent. Here the no-hole, one-hole, two-hole and so on, aspect of things determine their qualitative nature. Such theories recognize that many if not most interesting phenomena in nature involve discontinuities and catastrophe theory was designed to systematically categorize their types.

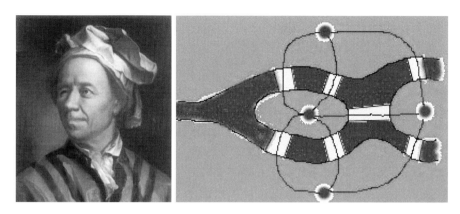

Figure 1.7: Leonhard Euler and the Seven Bridges of Königsberg problem. An Eulerian path through the bridges is possible only if there are exactly zero or two nodes of odd degree. In the latter case an Eulerian path must start and end at those two points. Since the graph corresponding to Königsberg has four nodes of odd degree (3 and 5), it cannot have an Eulerian path.

Leonhard Euler (1707-1783) was the first to use topological arguments to solve a curious problem known as the Seven Bridges of Königsberg, which is now a famous problem in introductory mathematics, and led to the branch of mathematics known as graph theory (see Figure 1.7). Graph theory has become increasingly popular over the past decade due to its utility in studying the properties of complex networks. Euler demonstrated that a route through the town of Königsberg (now Kaliningrad) that would cross each of its seven bridges exactly once did not exist. The solution is an example of topological argument because the result does not depend on the lengths of the bridges, nor on their distance from one another, but only on their connectivity properties, that is, its qualitative form. The above topological mathematics approach to physical problems stands in sharp contrast to the vast majority of previously available techniques that were designed for the quantitative study of continuous behavior.

The topological structure of a network determines in large part how calculations should be carried out to correctly describe its properties: adding or taking off one bridge in the town of Königsberg would greatly change the solution to the above problem. Thus, misrepresenting the real topological structure of a network might easily lead to a serious misunderstanding of that network. In fact, a complex network is determined not just by the fact that it is constituted by several components, but by how these components are connected together. It is evident that any network (biological, sociological, electronic, geophysical, etc.), which is commonly described as a complex system, is not just a random or confused aggregation of its components, but is a system characterized by its own structure and topology. Thus, to properly understand a complex network its topological structure should be correctly identified and modeled.

It is evident that even if one of the connections is missing the entire model might be topologically different from the physical phenomenon the model is supposed to represent. The mathematical algorithm adopted to describe the phenomenon would be erroneous and, ultimately, the interpretation of the result misleading. The mathematical difficulty is evident even with very simple networks. For example, if an electronic device is made of two resistors, R_1 and R_2, and we ask for the net resistance R_{tot}, it is well known that the correct answer can be given only after the topology of the device, that is, how the two resistors are connected (in series or in parallel) is given.[2]

Topology is only one example of a mathematical discipline whose application to science emphasizes the qualitative over the quantitative. Another important example is the bifurcation behavior of nonlinear dynamical equations. A bifurcation is a qualitative change in the solution to a dynamical equation controlled by varying a parameter. A mathematical example commonly studied is the logistic map. The logistic map is a simple quadratic equation depending on only one free parameter. Despite its simplicity the logistic map is an archetype of how complex, chaotic behavior arises from elementary nonlinear dynamical equations. Figure 1.8 shows that changing the parameter value generates a sequence of bifurcations in which the period of the solution doubles, doubles again and so on as the parameter value is increased. Thus, this behavior is an example of a period-doubling cascade. Eventually the solution winds up being irregular (chaotic) in time and as such has suggested a new paradigm for the unpredictable behavior of complex systems.

In fact, the logistic map was originally introduced as a demographic model in 1838 by Pierre François Verhulst (1804-1849) [175] and popularized in 1976

---

[2]If the two resistors are connected in series, the net resistance of the device is the sum of the two resistance: $R_{tot} = R_1 + R_2$. But if the two resistors are connected in parallel (that is, according to a different topology), the net resistance of the device is given by a completely different equation: $R_{tot} = (R_1^{-1} + R_2^{-1})^{-1}$.

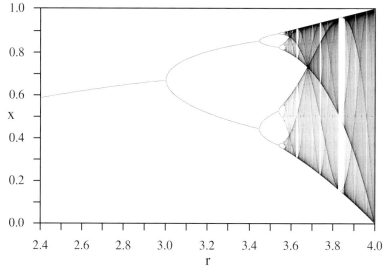

Figure 1.8: Pierre François Verhulst and the bifurcation diagram of the logistic map. $x_{n+1} = rx_n(1 - x_n)$ where r is the control parameter and $0 < x_n < 1$ evolves in time as $n = 1, 2, \ldots$ After a few interaction the value of x_n, given any initial values of $0 < x_0 < 1$, rapidly converges to the values plotted in the figure. For $0 < r < 1$, x_n converges to 0; for $1 < r < 3$, x_n converges to $(r - 1)/r$; for $3 \le r < 1 + \sqrt{6} \approx 3.45$, x_n oscillate between two values; then oscillates between 4 values, 8 values, 16 values *etc* becoming more and more chaotic with increasing r. For $r \ge 1 + \sqrt{8} \approx 3.83$, x_n oscillates between three values, and then 6 values, 12 values *etc*. For $r > 4$, x_n diverges.

in ecology by the biologist Robert May [101]. Its meaning refers to cases in which a population (represented by x_n, with $n = 1, 2, \ldots$ indicating the generation of the species) evolves according to two opposite causal effects: there is an increasing rate proportional to the current population (reproduction) in the presence of a limiting factor (starvation) that would decrease the population at a rate proportional to the complementary value of the current population measured in proportion to the capacity of the system. The complexity of the outcome depends significantly on the control parameter r representing a combined rate for reproduction and starvation as indicated in Figure 1.8.

Similar behavior is found in several natural phenomena. For example, the set of equations determining the air flow (wind) in the atmosphere is controlled by a parameter characterizing the strength of the coupling of the temperature of the air to the external forcing of the atmosphere. For certain values of the control parameter there exists convective rolls, which are harmonic solutions to the dynamic equations. The above properties of complex dynamical networks reinforces the common understanding of why although in specific situations we might predict the outcomes of some complex network, under different conditions our predictions fail even if the model equations are exactly the same. Thus, a particular quantitative solution to the model might be deceiving while true knowledge of the phenomenon may well be given by its qualitative properties.

This typical behavior occurs when the control parameter of a complex network falls in a range where the dynamics of the model becomes *chaotic*[3]. A long-term prediction fails in such a network because the microscopic errors in the specifications of the initial conditions and/or the computational round off errors in the calculation, are rapidly magnified from a microscopic scale up to a macroscopic scale. Figure 1.9 shows two solutions of the Lorenz system within the chaotic regime: even if the initial conditions are almost identical, after a few time units the two trajectories significantly depart from each other: a fact that clearly shows the impossibility of making long-term predictions for such a system. This effect is known as the *Butterfly Effect* and draws its name from the casual remark that was made about a lecture by Edward Norton Lorenz (1917-2008) whose title was: "Does the flap of a

---

[3]A nonlinear dynamical network is said to be chaotic if extremely small variations of the initial condition produce disproportionately large variations in the long-term behavior of the network.

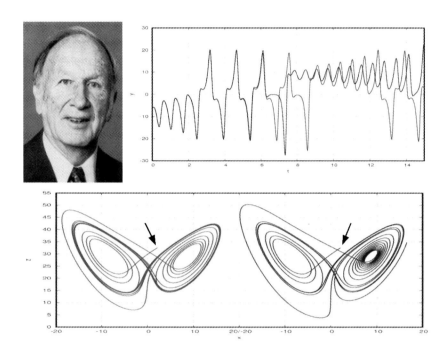

Figure 1.9: Edward N. Lorenz and the *Butterfly Effect*. In the example three coupled nonlinear equations are solved with slightly different initial conditions (0.1% difference only on $y(0)$) with the control parameter $r = 30$. The figure shows that after a few units of time the two curves significantly diverge although the initial conditions are almost identical. The equations, with the prime denoting a time derivative, are: $x' = 10(y-x)$; $y' = rx-y-zx$; $z' = -(8/3)z+xy$. The network represents a fluid diffusion model in which r is the imposed difference temperature between the bottom of the fluid layer and the top. When $r < 1$, the network is stable at $x = y = z = 0$, when $1 < r < 24.74$, the network has two stable regions with two centers, and when $r > 24.74$, the network behaves chaotically. The arrows show the position of the initial condition for the two networks indicated on the left and right.

butterfly's wings in Brazil set off a tornado in Texas?" This was mentioned in regard to a simple coupled network describing weather [96] that Lorenz was presenting at the 139th meeting of the American Association for the Advancement of Science in 1972.

1.2.2 What is a complex network?

A complex network is one of those entities that has either no description or a hundred descriptions, but no one description is complete. The best we can do is to construct a list of characteristics that a complex network ought to have; bearing in mind that such a list is always inadequate. First of all there are many elements (nodes) in the network, or many variables that are important to the network's development, most of which we do not know and have not measured. These elements could be the members of a particular organization such as a university or the military, the computers on the Internet, the ants in a colony, the bees in a hive, the neuron in the brain and so on. In any event, that development is determined by many relations among variables (mostly unknown), which typically, are described by coupled dynamical equations. The action potential propagating along an axon and generating the firing of other neurons in a network; the foraging patterns of ants; the cooperation of individuals in an organization; the pacemaker cells within the heart; all interact in different ways. The equations describing these interactions are generically nonlinear and subject to dynamical constraints imposed by the environment. The constitutive equations may be deterministic or random, continuous or discrete, but the *state of complexity* is typically characterized by a mixed qualitative structure where both *order* and *randomness* are present at the same time. This mixture insures a recognizable pattern, interpreted for example as long memory, in an apparently erratic process. Many of these concepts are condensed under the heading of scaling; a term we subsequently develop in a variety of contexts.

We need to identify the common features that exist across disciplines in order to obtain a working definition of complexity and complex networks. Let us assume that the solid curve in Figure 1.10 represent a measure of complexity. The mathematics of such systems near the bottom of the curve on the left, include integrable networks, used in celestial mechanics to predict the orbits of heavenly bodies. The equations of mechanics are determined by variations in the total energy of the system, which dictate how energy changes from kinetic to potential and back again, and in so doing restricts the pos-

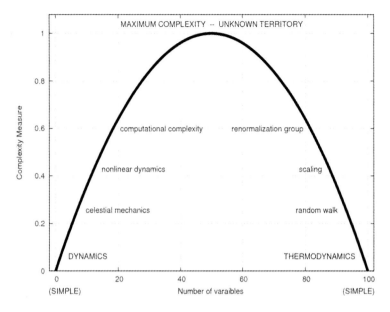

Figure 1.10: Here a conceptual nonlinear measure of complexity is depicted by the solid curve: one that categorizes networks with one or a few variables, described by deterministic trajectories, as being simple. In the same way networks that have a very large number of variables, not described by individual trajectories, but rather by probabilities, are also simple. Complexity lies between these two extremes of description having both the properties of randomness and regularity.

sible dynamics of the network. The smooth solutions to these equations are replaced in non-integrable networks, which describe nonlinear dynamical networks whose orbits break up into chaotic seas as the complexity increases. Further up the curve control theory is used to influence the dynamics of non-integrable networks through feedback loops and algorithmic complexity. The techniques describing the phenomena become more esoteric as the complexity rises with the increasing number of variables along the curve. But all these methods are relatively well understood while the phenomena are still comparatively simple. However, as complexity continues to increase with the increasing number of variables these techniques become less useful and blend into what we do not yet understand.

The mathematical techniques referred to in Figure 1.10, when used to formalize the context of a discipline, constitute what we mean by knowledge. For example, we know with absolute certainty that the sun will rise tomorrow or that the four seasons will follow each other. However, the reason is not just because it has done so for the past few billion years, although that might convince some people; but because Newton's law of universal gravitation and celestial mechanics predict that it ought to do so and we can theoretically deduce that the entire solar system because it is constituted by a single star is quite stable and the orbits of the planets quite predictable. Moreover, these predictions are consistent with experiments made on mechanical networks on the earth's surface and it is this consistency between the terrestrial and the celestial that gives us confidence in the prediction about the Sun and the orbit of the Earth. Of course this is only one kind of knowledge, which becomes diffuse when the deterministic dynamics become chaotic, as would happen in the case of a binary solar system where the orbits of the planets might follow complex paths because of the combined gravitational attraction from two different stars. Networks become less predictable and therefore less knowable as the complexity curve is ascended toward the maximum.

On the other side of the curve, when the number of variables describing the network is very large, we have equilibrium thermodynamics, which is based on random fluctuations, and the network is again apparently simple. The mathematics describing such networks can involve equations describing the evolution of probabilities to predict possible futures of the network; renormalization group relations that determine the general properties of such complex networks without solving equations of motion and scaling that determines the coupling of nodes across disparate space and time scales, all of which assists our understanding of the physical and life sciences. Here again as we ascend the curve, but now going in the backward direction where the network being modeled increases in complexity as the number of variables decreases, the stochastic mathematical tools available become less useful.

Again the mathematical techniques used in a disciplinary context to make predictions constitute what we mean by knowledge, but on the right-hand side of the curve that knowledge, which is based on a stochastic network topology, is very different from what we had on the left-hand side of the same curve, that is, in simple dynamical networks. For example, while in dynamical networks the experiments are easily reproducible, in the stochastic networks no experiment is exactly reproducible. Consequently, every experiment gives a different result and this ensemble of results is characterized by a probability density. It is this probability density that is used to predict the outcome of the next experiment in a succession of experiments. The uncertainty associated with the probability expands the notion of what constitutes knowledge beyond the certainty found in simple dynamical networks. The unknowability characteristic of a completely random network becomes knowable as the probability density takes on more deterministic characteristics such as scaling. From these observations we see that, like complexity, knowledge is neither a simple nor a single concept, but changes its forms according to whether we attempt to describe a phenomenon dynamically or stochastically.

The unknown territory of maximum complexity lies between the two extremes of simplicity; that being the trajectory of a cannon ball and the flipping of a coin. The area of maximum complexity is where we know the least scientifically and mathematically. It is where neither randomness nor determinism dominates, nonlinearity is everywhere, all interactions are nonlocal and nothing is ever completely forgotten. Here is where turbulence lurks, where the mysteries of neurophysiology take root, and the secrets of DNA are hidden. All the problems in the physical and life sciences that have for centuries confounded the best minds are here waiting for the next scientific/mathematical concept to provide some light. But let us not overlook the social and psychological sciences, the secrets of a harmonious society and human happiness reside in this veiled region as well.

1.2.3 A taxonomy of complex networks

It is always useful to approach difficult problems from a variety of perspectives. Generally, each view contributes another piece to the puzzle and a scientist is nothing if not a puzzle solver. One way to form a different view is to change the way information is organized and extracted. This organization is termed a *taxonomy*. Consequently, a different taxonomy of complex networks than the one presented based on dynamics in the previous section might be of value. Science is partitioned into various representative disciplines; starting with physics and the understanding of individual particles at the most basic level, blending into chemistry as particles interact with one another to form molecules and compounds, which morphs into biology as molecules order themselves into membranes to carry out functions and upwards to the social domain where the elements are individual people. At each level of this functional hierarchy there is a fundamental qualitative change in the form of the phenomenon under study. This change at each level makes it necessary to develop particular kinds of mathematics that would assist in the development of modeling and simulation of experimental results. Such mathematics would enable us to organize data into information. Information is then interpreted and this forms our knowledge of the phenomenon. However, at each level, sooner or later, the discipline encounters a barrier that cannot be surmounted using existing mathematical/scientific techniques. This is the barrier of complexity.

The descriptors of the phenomena of interest to scientists, administrators, healthcare givers, first responders, now and in the future, include multi-scale and multi-resolution analysis of nonlinear dynamic equations. In addition, the analysis of complexity supports the research to identify, understand and mathematically formulate the metrics in complex physical, biological and informational networks in order to enhance the detection, identification and response to threats to survivability, as well as to inhibitions of adaptiveness. For example, the early detection of local failures that may lead to catastrophic failure of the power grid, the Internet, air transportation and stock market are crucially important in the modern world. As a working definition we propose that complex adaptive networks have multiple interacting components, with behavior that cannot be simply inferred from the dynamics of the components, as it would be for a reductionistic description. The nonlinear interactions give rise to a blend of regular and erratic variability in complex networks. This variability may enable a network to behave in three different basic ways according to how its microscopic and macroscopic forms respond. In more complex networks these three behaviors might coexist to different degrees at multiple scales.

The first kind of behavior is encountered when a microscopically small change in a network greatly effects its macroscopic activity, as when a network's dynamics suddenly departs from its forecasted development. This behavior is observed, for example, in the chaotic activity of the Lorenz network because of its butterfly-effect sensitivity to microscopic variations (see Figure 1.9). A second type of behavior is encountered when both microscopic and macroscopic scales induce a major transformation as when the network suddenly switches into a different physical phase by reorganizing its internal structure and macroscopically becoming something qualitatively different: the bifurcation diagram of the logistic map or more commonly the freezing of water are typical examples. A third behavior is encountered when a significant microscopic change is required for leaving qualitatively unchanged the network at a macroscopic scale as when a network needs to adapt to a changing environment for survival. This later behavior occurs especially for biological and geophysical networks that evidently require a significant degree of robustness and malleability at the same time. Simple examples are: a) the mammalian thermostat where body temperature is kept constant by means of complicated microscopic biological mechanisms in response to external environmental changes; b) the Earth's thermostat where cooling allows CO_2 to build up in the atmosphere, while heating causes rain to reduce CO_2 in the atmosphere and so strengthens or weakens the *green house effect*, respectively; c) the solar thermostat where a decline in the Sun's core temperature causes the fusion rate to drop, so the Sun's core contracts and heats up while a rise in the sun's core temperature causes the fusion rate to rise, forcing the sun's core to expand and cool down.

Thus, if we do not know *a-priori* the characteristic of a given complex phenomenon we need to develop strategies for learning the properties of complex adaptive networks. One strategy commonly adopted is to perturb the network and record its response. The set of responses often constitutes the only information an observer can extract from a network. The difficulties encountered in understanding, controlling and/or predicting these responses are measures of complexity. In fact, complexity may be defined as that qualitative property of a model which makes it difficult to formulate its overall behavior, even when given reasonably complete information about its atomic components and their inter-relations. In everyday usage, networks with complicated and intricate features, having both the characteristics of randomness and order, are called complex. But, there is no consensus on what constitutes a good quantitative measure of complexity.

The most subtle concept entering into our discussion of complex adaptive networks up to this point is the existence and role of randomness. From one perspective the unpredictability associated with randomness has to do with the large number of elements in a network [100], but a large number of variables although sufficient, may not be necessary for the loss of predictability. Scientists now know that having only a few dynamical elements in the network does not insure predictability or knowability. It has been demonstrated that the irregular time series observed in such disciplines as economics, communications, chemical kinetics, physics, language, physiology, biology to name a few, are at least in part due to chaos [82]. Technically, chaos is a sensitive dependence on initial conditions of the solutions to a set of nonlinear, deterministic, dynamical equations. Practically, chaos

implies that the solutions to such equations look erratic and may pass all
the traditional tests for randomness, even though the equations are deterministic.[4] Computer programs are usually adopted to generate sequences of
numbers attempting to simulate intrinsically random physical processes, like
the elapsed time between clicks of a Geiger counter in proximity to a sample
of radioactive material.

Therefore, if we think of random time series as complex, then the output of a chaotic generator is complex. However, we know that something as
simple as a one-dimensional, quadratic map, the logistic map, can generate
a chaotic sequence. Using the traditional definition of complexity, it would
appear that chaos implies the generation of irreducible forms from simplicity. This is part of the Poincaré legacy of paradox [127]. Another part of
his legacy is the fact that chaos is a generic property of nonlinear dynamical networks, which is to say chaos is ubiquitous; all networks change over
time, and because they are nonlinear, they do, in principle, manifest chaotic
behavior for some parameter values.

A deterministic nonlinear network, with only a few dynamical variables,
can have chaotic solutions and therefore its dynamics can generate random
patterns. So we encounter the same restrictions on our ability to know and
understand a network when there are only a few dynamical elements as when
there are a great many dynamical elements, but for very different reasons.
Let us refer to the random process induced by the interaction of many variables as *noise*, the unpredictable influence of the environment on the network
of interest. Here the environment is assumed to have an infinite number of
elements, all of which we do not know, but they are coupled to the network of interest and perturb it in a random, that is, unknown, way [89].
By way of contrast chaos is a consequence of the nonlinear, deterministic
interactions in an isolated dynamical system, resulting in erratic behavior
of, at most, limited predictability. Chaos is an implicit property of nonlinear dynamical networks, whereas noise is a property of the environment
in contact with the network of interest. Chaos can therefore be controlled
and predicted over short time intervals, whereas noise can neither be predicted nor controlled, except perhaps through the way it interacts with the
network.

---

[4]Simple examples are found in the *pseudo-random* generators in computers whose simplest form is given by the *linear congruential generators* implemented in the ANSI C rand()
command: $I_{j+1} = aI_j + c \pmod{m}$ with $a = 1103515245$, $c = 12345$, and $m = 2^{32}$.

The above distinction between *chaos* and *noise* highlights one of the difficulties of formulating unambiguous measures of complex networks. Since noise cannot be predicted and only marginally controlled through filtering it might be viewed as being complex, thus, networks with many variables manifest randomness and may be considered complex. On the other hand, chaotic networks with only a few dynamical elements might be considered simple. In this way the idea of complexity is again ill posed, because very often we cannot distinguish between chaos and noise, so we cannot know if the network manifesting random behavior is formally simple or complex. Consequently, noise and chaos are often confused with one another and this fact suggests the need for novel approaches to the understanding of complex networks.

What we learn about complex networks is determined by the measurements we make, for example, forming time series from physical observables. In the example of the Earth's atmosphere the temperature at multiple locations on the surface and the resulting average global temperature obtained from those measurements form interesting time series for assessing the state of the Earth's climate. It should be pointed out that if a time series is used to make forecasts, the error in the forecast made by noise is significantly different from the error in the forecast made by chaos. This difference in the way error propagates in the two cases has been used to distinguish between chaos and noise [170].

In early papers on generalized systems theory it was argued that the increasing complexity of an evolving system can reach a threshold where the system is so complicated that it is impossible to follow the dynamics of the individual elements, see for example, Weaver [181]. Beyond this threshold new properties often emerge and the new organization undergoes a completely different type of dynamics. The details of the interactions among the individual elements are substantially less important, at this point, than is the *structure* (the geometrical pattern) of the new aggregate [33]. This is self-aggregating behavior; increasing the number of elements beyond this point, or alternatively increasing the number of relations among the existing elements, often leads to complete *disorganization* and the stochastic approach again becomes a viable description of the network behavior. If randomness (noise) is now considered as something simple, as it is intuitively, one has to seek a measure of complexity that increases initially as the number of variables increases, reaches a maximum where new properties may emerge, and eventually decreases in magnitude in the limit of the network having an infinite number of elements, where thermodynamics properly describes the network, cf. Figure 1.10.

The recent understanding of complex networks began with random networks, followed by small world theory, which was replaced with scale-free networks that opened the floodgates in the scientific literature over the past decade. The details of these various models of complex networks are discussed subsequently, but it is useful to have their general structure in mind in the following discussion. Each model emphasizes a different aspect of network complexity. The mathematicians Erdös and Rényi [46] in the 1950s and 1960s stripped away the properties distinguishing one complex network from another and considered the connections between structureless nodes. The property they identified as being important in these relatively simplex networks was whether or not two elements in the network were connected, and if these connections were made randomly what was the probability that a given node has a specific number of connections. Such a random network has properties that can be described by a simple one-humped function for the probability. The hump in the function occurs at the average number of connections, similar to the distribution curve large classes are graded on in college.

The next step in the progression of the development of a theory of complex networks came from the recognition that in real-world networks connections are not made at random. Watts and Strogatz [179] assumed that in social networks there are two distinct kinds of connections; strong and weak. The strong ties occur between family and friends and form clusters or cliques within a network. Then there are the weak ties; such as with many of the colleagues at work, friends of friends, and business acquaintances.. Clusters form among individuals having strong interactions, forming closely knit groups; clusters in which everyone knows everyone else. These clusters are formed from strong ties, but then the clusters are coupled to one another through weak social contacts. The weak ties provide contact from within one cluster to other clusters in the outside world. These weak connections are random and lead to small-world theory where with relatively few of these long-range random connections it is possible to link any two randomly chosen individuals with a relatively short path. This was the first theoretical explanation of the "six-degree-of-separation" phenomenon [180]. The two important properties of small-world theory are clustering and random long-range correlation, which along with other properties of the theory are discussed in Chapter 4.

Small-world theory is the conceptual precursor to understanding scale-free networks. Recent research into the study of how networks are formed and how they grow over time reveals that even the smallest preference introduced into the selection process has remarkable effects. Two mechanisms seem to be sufficient to obtain the scale-free distributions that are observed in the world. One of the mechanisms is contained in the principle that the rich get richer and the poor get poorer [8]. In a computer network context, this principle implies that the node with the greater number of connections attracts new links more strongly than do nodes with fewer connections, thereby providing a mechanism by which a network can grow as new nodes are added. The probability no longer has a single peak as it did for the random network but has a very long tail. Compare the distribution of heights with the distribution of income. The probability that the next person you meet on the street is approximately your height is very large, and the probability that the person is twice your height is almost certainly zero. On the other hand the probability that the next person you meet earns twice as much money as you is significant. It is even possible for that person to earn ten, or hundred or even a thousand times as much as you; depending, of course, on the street you are walking. These latter distributions are inverse power law and scale-free [9].

The inverse power-law nature of complex networks affords a single conceptual picture spanning scales from those in the World Wide Web to those within an organization. As more people are added to an organization the number of connections between existing members depends on how many links already exist. In this way the status of the oldest members, those that have had the most time to establish links, grows preferentially. Thus, some members of the organization have substantially more connections than do the average, many more than predicted by any single-humped distribution. These are the individuals out in the tail of the distribution, the gregarious individuals that seem to know everyone and to be enfolds in whatever is going on. Thus, the two factors, growth and preferential attachment, are responsible for the difference between random network models and real world networks [19].

There is increasing focus on the search for the most useful network topology. A conviction shared by a significant fraction of the scientists involved in research into the underlying properties of a science of networks is that there exists a close connection between network topology and network function. There is also increasing interest in network dynamics and its relation to network topology.

1.2.4   The three-tiers of science

A few decades ago physics, and most of science was neatly separated into theory and experiment.[5] One side of the physics community would testify that physics is an experimental science, as previously mentioned. When conflict would arise between observation and prediction, it is the controlled experiment that determines the truth of the matter. On the other side of the physics community are those who maintained that theory provides the guidance for new experiments as well as the context in which to interpret existing experiments. Consequently, they believed that physics is not strictly an experimental science, but is balanced against theory. Together they are two sides of the same coin, experiment and theory; data and knowledge. So where does information enter the picture?

The traditional two-tiered paradigm of science was revolutionized, that is disrupted, through the introduction of the programmable computer to solve complex mathematical models (see Figure 1.11). The theory that is too complex to yield predictions in the form of simple functions could be evaluated using the computer. By the same token, experiments that are too expensive to be carried out on a case by case basis, can be replaced or augmented by computer simulations. Consequently, the computer extends science from both the side of experiment through simulation and the side of theory through evaluation. This intermediate zone is the domain of information, which forms the third tier of today's three-tier construction of science. The unfortunate aspect of this observation is that it is not always shared by the general scientific community. In fact, many practitioners of the traditional scientific disciplines proceed with an experiment/theory view that actually inhibits understanding complex adaptive networks. From the restricted two-tier perspective the computer is used primarily as a computational adjunct to the two historical domains. Although virtually every university has substantial investment in computers and computation, the view of science has not expanded beyond the theory/experiment paradigm except at the very top level research schools.

One of the pioneers of this scientific revolution was the polymath John von Neumann (1903-1957). He made significant contributions to many fields of science and contributed to the mathematical foundations and interpretation of measurement theory in quantum mechanics and laid the groundwork for

---

[5]We take up the issue of experiment in the chapter on data and that of theory in the chapter on knowledge.

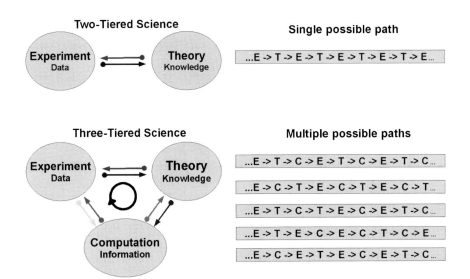

Figure 1.11: Comparison between a simple two-tiered science and a three-tiered science. Note that the two kinds of science are topologically distinct and, therefore, fundamentally different. In fact, a two-tiered science has just two connected nodes and this forces a scientific investigation to follow a unique topological path in which experiments are used to improve theory and the theory is used to improve experiments in an alternating pattern. On the contrary a three-tiered science has three interconnected nodes that give rise to a complex network linking of experiment, theory and computational information. The increased complexity of the network is evident because a scientific investigation can now follow different topological paths that are all equally permissible, that is, compatible with the scientific method. However, these different paths do not commute and one path might yield a scientific investigation that is closer to the truth more rapidly than another.

Figure 1.12: John von Neumann is considered the co-creator of game theory, of the concepts of cellular automata and the universal constructor and contributed to the development of the Monte Carlo method, which allowed complicated problems to be approximated using random numbers. He is the father of the single-memory computer architecture in which data and program memory are mapped into the same address space. This architecture, with very few exceptions, is used in all present-day home computers, microcomputers, minicomputers and mainframe computers. Computers have driven the evolution of science from a simple two-tiered network to a complex three-tiered network.

the design of programmable computers. He also created the subject of Games Theory in 1928 [1]. Games Theory consisted of proving that a quantitative mathematical model could be constructed for determining the best strategy, that is the one which, in the long term, produces the optimal result with minimal losses for any game, even one of chance or one with more than two players. The sort of games for which this theory found immediate use were business, warfare and the social sciences — games in which strategies must be worked out for defeating an adversary.

However, most of von Neumann's career was spent at the *Institute for Advanced Study* where he designed and supervised the construction of the first computer able to use a flexible stored program (MANIAC-1) in 1952. His interest in developing the computer was, in part, to enable physical scientists to calculate predictions for complex networks such as the weather. This interest was fueled by his World War Two research into hydrodynamics, shock wave propagation and the H-bomb. He was confident that the computer would enable scientists to overcome the mathematical barriers to obtaining solutions to the equations for complex adaptive networks by skirting the lack of mathematical solutions and going directly to calculations without the intermediate step of solving the nonlinear equations.

Let us consider, for example, the large-scale numerical calculation associated with turbulent fluid flow, say the white water churning through the Grand Canyon. Without recounting the myriad of technical difficulties associated with solving the problem of turbulence it is worth pointing out that there is a million dollar prize offered by the Clay Institute [39] to the person who can prove the existence of a general solution to the underlying equations of viscous fluid flow; the Navier-Stokes equations (see Figure 1.13). In fact, these equations must be solved in order to calculate the response of an aircraft to various atmospheric conditions, including clear air turbulence; the flow of air into the human lungs; the dispersal of oxygen throughout the body by the blood taking up the oxygen from the air in the lungs, as well as almost any phenomena involving fluid flow in the real world.

We observe that large-scale numerical calculations generate results that are often no easier to understand than are the outcomes of the experiments they are intended to simulate. It is necessary to develop simple models for pieces of the experiment that enable the researcher to break the calculation into manageable conceptual parts. In turbulent flow, such a piece might be the water flow near a boundary to understand the transition from turbulent flow adjacent to the boundary to smooth flow some distance from the wall. These parts often constitute components of the network that an engineer can design and in so doing understand how that component contributes to the larger network dynamics. Evidently two theoretical models with different pieces of information might have very different outcomes and yield two very different understandings of a physical phenomenon. For example, it was not until Harvey identified the human heart as a pump, that positive intervention in the cardiovascular network could be made.

Evidently, these remarks do not refer only to the *hard* sciences, such as physics, chemistry and biology. Indeed, the two-tiered paradigm of science has been used historically to try and develop also the *soft* sciences such as sociology and psychology and understand the phenomena they address.

Navier-Strokes Equations
3-dimensional - unsteady

Coordinates: (x, y, z) Time: t Density: ρ Pressure: p Reynolds Number: Re

Velocity Components: (u, v, w) Stress: τ Heat Flux: q Prandtl Number: Pr

$$Continuity \;:\; \frac{\partial \rho}{\partial t} + \frac{\partial(\rho u)}{\partial x} + \frac{\partial(\rho v)}{\partial y} + \frac{\partial(\rho w)}{\partial z} = 0$$

$$X - Momentum \;:\; \frac{\partial(\rho u)}{\partial t} + \frac{\partial(\rho u^2)}{\partial x} + \frac{\partial(\rho uv)}{\partial y} + \frac{\partial(\rho uw)}{\partial z} = -\frac{\partial p}{\partial x} + \frac{1}{Re_r}\left(\frac{\tau_{xx}}{x} + \frac{\tau_{xy}}{y} + \frac{\tau_{xz}}{z}\right)$$

$$Y - Momentum \;:\; \frac{\partial(\rho v)}{\partial t} + \frac{\partial(\rho uv)}{\partial x} + \frac{\partial(\rho v^2)}{\partial y} + \frac{\partial(\rho vw)}{\partial z} = -\frac{\partial p}{\partial y} + \frac{1}{Re_r}\left(\frac{\tau_{xy}}{x} + \frac{\tau_{yy}}{y} + \frac{\tau_{yz}}{z}\right)$$

$$Z - Momentum \;:\; \frac{\partial(\rho w)}{\partial t} + \frac{\partial(\rho uw)}{\partial x} + \frac{\partial(\rho vw)}{\partial y} + \frac{\partial(\rho w^2)}{\partial z} = -\frac{\partial p}{\partial z} + \frac{1}{Re_r}\left(\frac{\tau_{xz}}{x} + \frac{\tau_{yz}}{y} + \frac{\tau_{zz}}{z}\right)$$

$$TotalEnergy - Et \;:\; \frac{\partial(E_r)}{\partial t} + \frac{\partial(uE_r)}{\partial x} + \frac{\partial(vE_r)}{\partial y} + \frac{\partial(wE_r)}{\partial z} = -\frac{\partial(up)}{\partial x} - \frac{\partial(vp)}{\partial y} - \frac{\partial(wp)}{\partial z}$$

$$+ \frac{1}{Re_r}\left[\frac{(u\tau_{xx} + v\tau_{xy} + w\tau_{xz})}{x} + \frac{(u\tau_{xy} + v\tau_{yy} + w\tau_{yz})}{y} + \frac{(u\tau_{xz} + v\tau_{yz} + w\tau_{zz})}{z}\right]$$

$$- \frac{1}{Re_r Pr_r}\left(\frac{\partial q_x}{\partial x} + \frac{\partial q_y}{\partial y} + \frac{\partial q_y}{\partial z}\right)$$

Figure 1.13: The three-dimensional unsteady form of the Navier-Stokes Equations. These equations describe how the velocity, pressure, temperature, and density of a moving viscous fluid are related. The equations were derived independently by G.G. Stokes, in England, and M. Navier, in France, in the early 1800's.

However, this has not been very successful. In fact, the notions of network elements, relations, interactions, emerging properties, nonlinearity and so on, do not depend on the network being physical, chemical or biological. These ingredients also go into psychological and social networks, without significant change. It appears that what made the *hard* sciences amenable to the mathematics of the nineteenth century and the success of the two-tiered approach was the simplicity of the phenomena being studied. Much of the physical world, in the narrow range of parameter values necessary for human survival, is made up of linear additive processes and consequently the linear mathematics of waves, diffusion and harmonic motion are often adequate for their description.

However, when a wave amplitude becomes too large, or the changes in the diffusing chemical concentrations become too sharp, or the oscillating material exceeds its elastic limit, linear mathematics no longer applies and must be replaced with more accurate nonlinear descriptions. In some physical networks the dynamics become nonlinear, in others the interactions become nonlocal in space and acquire a dependence on the network's history. In this respect the physical network, when operating outside *normal* regions, behaves more like the typical operations within the soft sciences. Thus, the soft sciences almost always consist of complex adaptive networks that are nonlinear, stochastic and multiplicative, whereas this is only sometimes true of the hard sciences.

This explains, at least in part, why the methods used so successfully in the hard sciences have had only minor successes in the soft sciences. The two-tiered science was inadequate for the soft sciences, which, because of the inherent complexity of the phenomena, required the introduction of the third scientific tier, computation, in order to organize experimental data into information and to systematically interpret that information in terms of theory.

The difference between the two kind of scientific investigations is not superficial, but substantial. As Figure 1.11 shows a two-tiered science and a three-tiered science are topologically different. While a two-tiered science might develop iteratively according to a predictably unique path, the possibility of linking three nodes — experiment, theory and computational information — yield to the formation of a complex network where a scientific investigation might iteratively move along different paths. For example, new experimental data might be used to update theory or, alternatively, the computational algorithms; a new theory might suggest conducting new experiments or, alternatively, new computations and, finally, new computations might suggest new experiments or, alternatively, formulate new theory, and so on. However, although these paths are all equally permissible according to the scientific method, they are not expected to be interchangeable and consequently might lead to different "scientific truths" concerning the same phenomena at any given point in time.

Thus, one specific path, or scientific strategy, is better than another in the sense that it is more efficient for disclosing the *objective* truth about a phenomena. This means that an investigator should pre-determine the optimum path and/or, if this is not possible, s/he should continuously enquire into the limitations of following a specific path, and/or acknowledge that there might be different ways to address a given phenomenon. These different paths might also lead to different conclusions which are both scientifically valid. By not acknowledging the existence of multiple investigative paths, a researcher might unconsciously introduce into his/her scientific enquiry a subjective component that could result in misleading conclusions about the nature of the phenomenon being investigated. Such mistakes arise because the scientific investigation was conducted not via the most appropriate scientific path but via a path that was just the most familiar to the investigator.

We now try to clarify these ideas by examining the concrete physical problem of global warming. In particular, we discuss the nature of data and the importance of experiment and/or observation in science. Assessing the patterns found in the data and their interpretation in terms of theory is taken up in the later chapters.

1.3   Framing the climate change debate

Global warming is one of the most debated issues in contemporary science. The phenomenon refers to the fact that the observed global average air temperature near the Earth's surface rose about 0.8 °C during the last 100 years as shown in Figure 1.14. The debate is *not* about the rise in average global temperature, although the exact magnitude of this rise is still debated [107], but about what caused this rise.

For better or for worse the global warming debate is not restricted to the scientific community, but has expanded to the general public and has deeply involved reporters, social activists and politicians from a significant number of countries. On any given day one can find several newspapers articles and web-sites passionately disputing the pros and cons of global warming in terms that are not strictly scientific, but are frequently driven by emotions and/or social-political-economic positions. The positions of the antagonists range from the catastrophic alarmists, who foresee the eminent demise of the Earth, to those who deny the very existence of global warming. The authors are at neither extreme and we restrict our discussion to the science, avoiding politics whenever possible. However, the preliminary discussion of the importance of the problem would not be complete without some commentary on the political debate.

The ongoing world-scale political and public debate on global warming is understandable considering that a major climate change would have

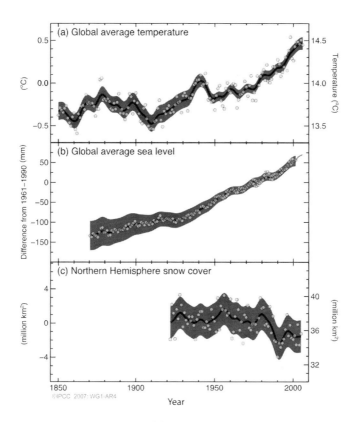

Figure 1.14: Observed changes in (a) global average surface temperature, (b) global average sea level from tide gauge (blue) and satellite (red) data and (c) Northern Hemisphere snow cover for March-April. All changes are relative to corresponding averages for the period 1961-1990. Smoothed curves represent decadal average values while circles show yearly values. The shaded areas are the uncertainty intervals estimated from a comprehensive analysis of known uncertainties (a and b) and from the time series (c).

dramatic consequences for humanity. Among the most serious effects of global warming discussed by various scientists are: the melting of the glaciers that will cause the oceans' level to rise with the evident outcome that large amounts of water will overflow onto the land forcing several hundred million people living on the coasts to migrate inland; an increase of the extension of the deserts that would again result in forced migration for an estimated hundred million people; the ocean acidification induced by an increased atmospheric CO_2 concentration that might alter the sensitive marine ecosystem with consequent disruption of food webs and, therefore, impacting the human societies that depend on them. Other predicted consequences include the increased intensity of extreme weather events such as tornados, changes in the amounts and patterns of precipitation, with serious consequences for agricultural activity, the extinctions of several species of animals, in particular polar bears, and plants and a wide increases in the ranges of disease vectors. All this would have evident catastrophic humanitarian consequences, specifically in the poorest and more densely populated regions of the Earth.

Thus, not only scientists but the larger community, from national governments to small circles of friends drinking beer in a pub, can be found engaged in discussing what, if any, action should be taken to reduce or reverse future warming. Up until now a large number of nations have signed and ratified agreements such as the Kyoto Protocol, aimed at reducing anthropogenic greenhouse gas emissions, which are considered responsible for the observed global warming. Other nations, such as the United States, which is currently producing the highest concentration of emissions, have refused to sign the Kyoto Protocol. The refusal is the result of serious political-economic disagreement on the degree of the exceptions requested by some newly-developed economies, such as China and India, which are expected to exceed the U.S. emissions within the next few years.

However, sidestepping the political-economic-sociologic aspects of the global warming debate, herein we focus on the science. It is evident that climate is a complex network composed of innumerable elements. The scientific challenge is to identify and separate the various mechanisms that contribute to this network and estimate the relative importance of their individual contributions. The intrinsic complexity of climate and the significant data uncertainty has made this issue more and more controversial during the last few years. In the sequel we briefly discuss this point and frame the global warming debate within a complex network scientific perspective.

1.3.1   The Intergovernmental Panel on Climate Change

The United Nations acknowledges the importance of evaluating the risks of climate change caused by humans themselves and attempts to involve all nations in a communal effort to address this important issue. In 1988 two United Nations organizations, the World Meteorological Organization (WMO) and the United Nations Environment Programme (UNEP) established the Intergovernmental Panel on Climate Change (IPCC) for the purpose of publishing special reports on topics relevant to the implementation of the United Nations Framework Convention on Climate Change (UNFCCC).

The principles governing the IPCC reports are based on the following statement:

> The role of the IPCC is to assess on a comprehensive, objective, open and transparent basis the scientific, technical and socio-economic information relevant to understanding the scientific basis of risk of human-induced climate change, its potential impacts and options for adaptation and mitigation. IPCC reports should be neutral with respect to policy, although they may need to deal objectively with scientific, technical and socio-economic factors relevant to the application of particular policies.

> Review is an essential part of the IPCC process. Since the IPCC is an intergovernmental body, review of IPCC documents should involve both peer review by experts and review by governments.

Thus, the IPCC reports are based on peer reviewed and published scientific literature and are prepared by several scientists working on related issues from several nations, but under the political supervision of the United Nations. Some fear that these works are excessively alarmist because a restricted group of scientists select the appropriate scientific literature and may use this occasion to advance their personal scientific and political agendas and, therefore, be less than objective. Others believe that the IPCC reports are very conservative and that the real condition of climate is much worse than what they depict because of the mediations of the various governments through their policy makers who have the final say in what goes into the IPCC reports.

In general, the IPCC assessments of climate change are the product of extensive scientific research and political negotiation and, although they can stimulate serious debate and controversy, they can also be considered a balanced summary of the current scientific/political understanding of the issue. Therefore, both the political and the scientific communities consider these reports to be authoritative. Of course science is based on objective facts (data), not on authority or popularity, and it cannot be excluded that future scientific developments shifts to more or less radical conclusions from those presently held. But we discuss this issue later in the book.

The IPCC has published four major assessment reports with supplements. These reports were published in 1990, 1995 (SAR), 2001 (TAR) and the most recent one was released in 2007 (AR4). Each assessment was prepared by three different Working Groups (WG) that prepared three different reports. For AR4, Working Group I dealt with the "Physical Science Basis of Climate Change," Working Group II dealt with the "Impacts, Adaptation and Vulnerability," and Working Group III dealt with the "Mitigation of Climate Change." Each report was preceded by a "Summary for Policymakers."

The summaries for policymakers were released before the reports and certain aspects of the summaries received a high level of visibility in the media. Many scientists find this procedure surprising and disturbing. In fact, the expectation by scientists is that a summary ought to faithfully conform to the actual report, as prepared by the scientists. Thus, the report should be available *before* the summary and not the other way around. However, according to a document entitled "Procedures for the preparation, review, acceptance, adoption, approval and publication of IPCC reports":[6]

> The content of the authored chapters is the responsibility of the Lead Authors, subject to Working Group or Panel acceptance. Changes (other than grammatical or minor editorial changes) made after acceptance by the Working Group or the Panel shall be those necessary to ensure consistency with the Summary for Policymakers or the Overview Chapter.

The above quote suggests that the final IPCC Reports are made *consistent* with the Summary for Policymakers, and not vice versa, as many would expect. This procedure suggests that politics plays a significant role in the conclusions reached in these documents. The notion of revising a report to conform to a policy summary is anathema to science and takes the work out of the scientific realm, and places it squarely in the political one. Indeed, IPCC

---

[6]See the web-page http://www.ipcc.ch/about/app-a.pdf.

reports serve the political more than the scientific community. Therefore, the reports seem to be more political than scientific documents.

However, it is worth reemphasizing that the way scientists look at a scientific issue is different than how a politician or media person looks at the same issue. Scientists know that the frontiers of scientific knowledge are filled with uncertainties. They are aware of the controversies surrounding any specific issue. For example, a scientist focuses attention not on what is understood, but on what is not understood. S/he concentrates on the details, and tries to emphasize the existence of possible scientific misconceptions. By doing so, s/he loves to continuously suggest new hypotheses, theories, calculations and experiments hoping to succeed in better understanding a specific phenomenon. Scientists are essentially *researchers*. What they are supposed to do is to humbly question Nature[7] in the hope of understanding *how God has made it*, as Galilei, Newton and Einstein would have phrased it.

The picture that the public and, therefore, the media and politicians have of a scientist, can be quite different from what they are in reality. In our modern culture a scientist is not necessarily perceived as a researcher, that is, as an individual seeking the truth concerning a specific phenomenon. A scientist is often expected to be a kind of *prophet*; one who is supposed to already know the *truth* about whatever topic s/he addresses. As any working scientist knows, nothing could be further from reality.

Indeed, politics ask for *certainty* and the media sells *extravaganza*. Thus, the subtle hypotheses, assumptions and details that often constitute the foundation on which the scientific explanation of a phenomenon is based are not just ignored, but are often attacked through sneering and innuendo. In this ignorance in particular those scientists that question some popular theory may not be understood and, consequently, they may not be considered relevant, and often they are also omitted or even censored in public debates. Scientists who try to explain the difficulties relating to a certain topic are easily considered *obscure* and dismissed. Indeed, some scientists might even take advantage of this situation and gain popularity by playing the role of prophet by proclaiming what the media would expect from them!

However, scientists are not prophets, their job consists in investigating nature, and the truth is that the frontiers of scientific knowledge always

---

[7]The scientist is humble in the sense that the answer to a scientific inquiry is not presumed to be known in advance. However, this does not mean that the questioner is shy in asking questions; to the contrary, scientific questioning is often very aggressive.

ferments debate and controversy. The hypotheses, the findings and their interpretation are questioned and evaluated again and again in a turbulent iterative process that is called the *scientific method*. It is understandable that non-scientists might be confused and distressed by contradictory scientific statements emanating from the scientific community, but this is how knowledge is forged; not by gentle smooth agreement, but often by strident controversy. This is how science has progressed through the centuries: by means of debate, controversy and disagreement. When phenomena are complex it is evident that debate and controversy will arise even more than what occurred in the past once it is clear that the three-tiered science allows a scientist to follow alternative investigative paths.

Thus, due to the intrinsic complexity of climate, it is not surprising that when the audience looks behind the curtain, what they find is that scientists are still debating many issues related to global warming. It would not be surprising if an actual legitimate understanding of the phenomenon today might be contradicted by another equally legitimate understanding of the same phenomenon tomorrow. These alternative paths will eventually converge only by means of a further deepening of climate science . Right now the scientific investigations still need to move on parallel alternative paths. But for the moment let us summarize what are the actual data concerning global warming as reported by IPCC.

The IPCC AR4 as presented by Working Group I that deals with the "Physical Science Basis of Climate Change" concludes:

> Most of the observed increase in globally averaged temperatures since the mid-20th century is very likely due to the observed increase in anthropogenic greenhouse gas concentrations.

Thus, natural phenomena such as solar variation combined with volcanoes have probably had only a small warming effect from pre-industrial times to 1950. Indeed, after 1950 natural forcing is stated to have had but a small cooling effect. At least 30 scientific societies and academies of science, including all of the national academies of science of the major industrialized countries have endorsed the above basic conclusions.

The calculations upon which the above conclusions are based have been made using general circulation climate models, which are large-scale computer codes, that take into account a number of nonlinear hydrodynamic and thermodynamic climate mechanisms and use total solar irradiance and

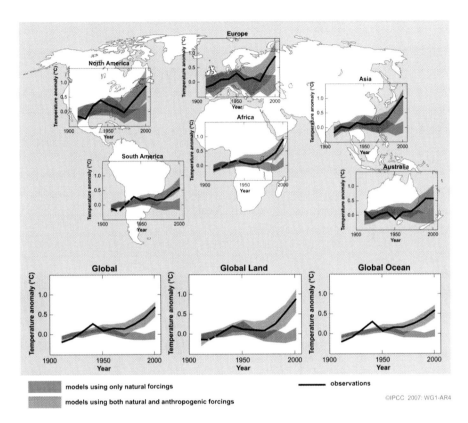

Figure 1.15: Comparison of observed continental and global scale changes in surface temperature with results simulated by climate models using natural and anthropogenic forcings. Decadal averages of observations are shown for the period 1906 to 2005 (black line) plotted against the center of the decade and relative to the corresponding average for 1901-1950. Lines are dashed where spatial coverage is less than 50%. Blue shaded bands show the 5-95% range for 19 simulations from five climate models using only the natural forcings due to solar activity and volcanoes. Red shaded bands show the 5-95% range for 58 simulations from 14 climate models using both natural and anthropogenic forcings [from [162] with permission].

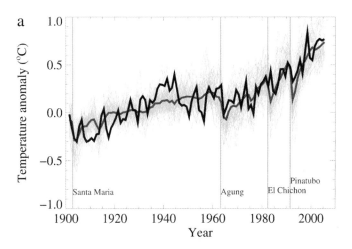

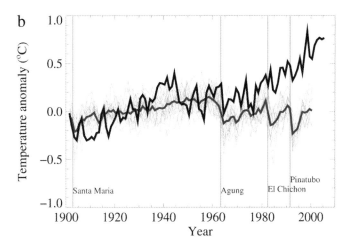

Figure 1.16: The figure shows a comparison of observed global surface temperature anomalies relative to the period 1901 to 1950 (black curve) with several climate model simulations using natural and anthropogenic forcings. When all (anthropogenic and natural) forcings are considered the model simulations, as obtained from 58 simulations produced by 14 models (red curve in the top figure), overlap the data quite well. However, when only natural (solar and volcano) forcing are considered the simulations, as obtained from 19 simulations produced by 5 models (blue curve in the bottom figure), significantly depart from the data since 1950 and show a cooling [from [162] with permission].

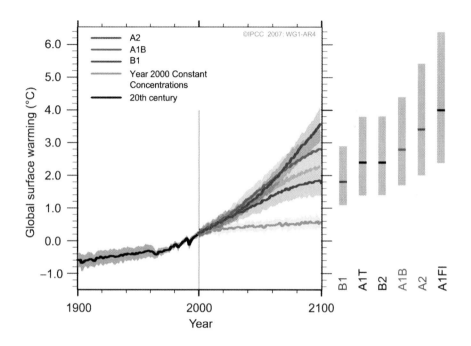

Figure 1.17: Solid lines are multi-model global averages of surface warming (relative to 1980-1999) for the scenarios A2, A1B and B1, shown as continuations of the twentieth century simulations. Shading denotes the ±1 standard deviation range of individual model annual averages. The orange line is for the numerical experiment where concentrations were held constant at year 2000 values. The grey bars at right indicate the best estimate (solid line within each bar) and the likely range assessed for the six SRES marker scenarios. The assessment of the best estimate and likely ranges in the grey bars includes the AOGCMs in the left part of the figure, as well as results from a hierarchy of independent models and observational constraints [from [162] with permission].

several terrestrial natural and anthropogenic forcing as input. Several climate simulations have been run and the results were compared with the local and global temperature data: see Figure 1.15. The figure shows an agreement between the data and the simulations when all estimated climate forcing are used in the sense that the climate simulations reproduce a warming. However, when the climate models are forced only with the natural forcing (that is the solar and volcano forcing) the simulations significantly depart from the data after 1950 by showing a cooling trend while the data show an abrupt temperature increase: see Figure 1.16. Thus, the logical conclusion would be that climate forcing associate with anthropogenic greenhouse gas concentrations is the one that has been responsible for the observed increase in globally averaged surface temperatures since the mid-20th century (confidence level >90%). In fact, global atmospheric concentrations of carbon dioxide, methane, and nitrous oxide have increased markedly as a result of human activities since 1750 and during the last decades far exceed pre-industrial values over the past 650,000 years. The probability that global warming is caused by natural climatic processes alone is estimated in the IPCC report to be less than 5%.

The IPCC report also includes predictions of climate change under several forcing scenarios for the 21st century. It is estimated that during the 21st century world temperatures could rise by between 1.1 and 6.4 °C (2.0 and 11.5 °F) (see Figure 1.17) and that sea levels could probably rise by 18 to 59 cm (7.08 to 23.22 in). Warm spells, heat waves and heavy rainfall will be more frequent (confidence level >90%) and there is a confidence level of >66% that there will be an increase in droughts, tropical cyclones and extreme high tides. Both past and future anthropogenic carbon dioxide emissions will continue to contribute to warming and sea level rise for more than a millennium.

The above conclusions represent the scientific understanding on which the other two IPCC reports, prepared by the Working Group II and III, are based and, therefore, depend. We do not discuss these other two reports herein. Most policymakers and the lay public would just trust the scientific findings presented in the report prepared by WG1 and look at the content of the reports prepared by WG2 and WG3, which deal with the "Impacts, Adaptation and Vulnerability," and the "Mitigation of Climate Change," respectively.

However, an authentic scientific investigation requires that the physical science basis be carefully evaluated. A scientist wonders whether the scientific framework on which the scientific interpretation of climate change is robust or if it may be flawed in some fundamental way. It is evident that there may be a significant number of small errors in both the data and the models, but this uncertainty would not be really important; the errors could eventually compensate for one another and the overall interpretation presented by the IPCC report would then not significantly change. If there is a major problem it has to do with the overall philosophical interpretation of the results; a problem that might result from associating a mistaken topology with the climate network.

1.3.2 Climate topologies in comparison

As we discussed in the previous paragraphs, associating the appropriate topology with a physical network is essential for the faithful interpretation of the network. Different topologies imply different qualitative properties of a network and these in turn imply different physical structures. Different topologies can lead to drastically different interpretations of the same *data* and suggest different ways to carry out and interpret calculations. The latter constitute the *information* about a network that we extract from the data. Ultimately, different topologies lead to different kinds of understanding or potential *knowledge* regarding a network.

All major scientific switches have been induced by some kind of changes in perspective in how reality is seen. The Copernican revolution, by its putting the Sun at the center of the solar system was a change in perspective from the Aristotelian/Ptolematic system that put an immobile Earth at the center. Note that the mathematical details of the Copernican system, with its perfect cycles and hemicycles, made it neither a better nor a simpler system than the geocentric one. However its implicit perspective allowed Kepler and subsequently Newton to correctly understand the motion of the planets and to discover the universal law of gravitation. Analogously, special and general relativity are the result of major changes in perspective. Specifically, Einstein had the intuition to switch from a model in which time is absolute to one in which the speed of light is absolute. Similarly, the passage from classical to quantum mechanics is characterized by a major change in perspective required for switching from a continuous deterministic understanding of the physical world to a discrete probabilistic one.

These ideas suggest that the inclusion of a new mechanism or form of interaction adds a new link to the network that in turn may lead to a new model with significantly different properties. Thus, determining the correct topology of a phenomenon, including the Sun-Earth climate network, is extremely important. If a topological error is found in the scientific basis underlying the IPCC report, a major change in the understanding of climate change may result. Herein, we investigate this possibility.

One of the main mechanisms underlying climate change is the *greenhouse effect*. This important climate mechanism was discovered by Joseph Fourier in 1829 and then better explained by Svante August Arrhenius in 1896 who developed a theory to explain that ice ages are caused by the changes in the levels of carbon dioxide (CO_2) in the atmosphere through the greenhouse effect [6].

The greenhouse effect is the process in which the trapping of infrared radiation by the atmosphere warms a planet's surface. The name comes from an analogy with the warming of air inside common greenhouses as used in agriculture compared to the air outside them. The mechanism regulating the heat of greenhouses is simple, visible light passes through the transparent glasses of the greenhouse and is absorbed by the darker surface of the ground and plants. These become warm and, as a result, the air inside the greenhouses also warms. At this point the warm air is prevented from rising and flowing away because greenhouses are closed structures. The air inside warms as a consequence of the suppression of convection and turbulent mixing. Thus, the glass walls and roof of a greenhouse allow most of the Sun's light in, but does not allow most of the heat to escape.

The Earth's atmosphere works in a similar, but not exactly in the same way. Greenhouses warm by preventing convection, that is, hot air cannot escape; the atmosphere's greenhouse effect however reduces radiation loss, not convection. This happens because the so-called *greenhouse gases* prevent the infrared radiation from escaping the atmosphere after it warms up.

The most important greenhouse gases are: water vapor (H_2O), which causes at least 60% of the greenhouse effect on Earth; carbon dioxide (CO_2), which produces 9-26% of the effect; methane (CH_4), which causes 4-9%; and ozone (O_3), which produces 3-7%. Other greenhouse gases include, but are not limited to, nitrous oxide (N_2O), sulfur hexafluoride (SF_6), hydrofluorocarbons ($HFCs$), perfluorocarbons ($PFCs$) and chlorofluorocarbons ($CFCs$). Note that major atmospheric constituents (nitrogen, N_2 and oxygen, O_2) are not greenhouse gases because homonuclear diatomic molecules neither absorb nor emit infrared radiation, so these gases are transparent to both visible light and infrared radiation.

Figure 1.18: Joseph Fourier (1768-1830) and Svante August Arrhenius (1859-1927). Fourier is credited with having understood that gases in the atmosphere trap the heat received from the Sun and consequently warm the atmosphere: this effect is today called the *greenhouse effect* for its analogy with the warming of air inside a greenhouse compared to the air outside it. Arrhenius further developed the ideas of Fourier and after a laborious numerical computation concluded that cutting the amount of CO_2 in the atmosphere by half could lower the temperature in Europe some 4-5 °C (roughly 7-9 °F) — that is, to an ice age level. If the CO_2 concentration in the atmosphere doubles, Arrhenius estimated it would raise the Earth's temperature some 5-6 °C. Curiously Arrhenius was more interested in explaining the ice ages and evidently thought that these were induced by large *natural* greenhouse gas changes. Arrhenius never thought that an *anthropogenic* global warming could ever be realistic. In fact, the CO_2 released from the burning of coal in the year 1896 would have raised the CO_2 level in the atmosphere by scarcely a thousandth part and it could have taken more than three thousand years for the CO_2 level to rise to such a level.

Clouds also contribute to the greenhouse effect but their role is only poorly understood. Different types of clouds have different effects and cloud are poorly modeled. This makes the total estimate uncertain. In fact, clouds can both increase the albedo of the Earth by reflecting sunlight (and therefore reduce the incoming energy that can be converted to heat) and intercept heat radiation from the Earth's surface and atmosphere and radiate heat back down, warming the surface. Thus, clouds both cool and warm the surface, and it is not well understood which of the two processes dominates: a fact that is determined by different circumstances and specific cloud properties. It is worth noticing that the current large uncertainty about the effect of clouds on climate is quite important because it prevents a precise calculation of the effect of the other greenhouse gases. According the IPCC a doubling of CO_2 may induce a temperature increase from 1.5 K to 4.5 K, and more. This is, indeed, a very large uncertainty!

In any case, the greenhouse gases are approximately transparent to visible light, which consequently passes through the atmosphere and reaches the Earth's surface. At the surface this light is absorbed and the energy remitted as infrared radiation. But now the greenhouse gases are opaque to the long wavelength infrared radiation and do not allow it to pass through and escape into space. The absorption of this longwave infrared radiant energy warms the atmosphere more than what it would be warmed by the transfer alone of sensible and latent heat from the surface, as calculated using the Stefan-Boltzmann law of radiation. The final result is that the Earth's average surface temperature is 292-293 degrees Kelvin or 14-15 °C which is about 33 K warmer than what it would be without the surplus warming effect caused by the greenhouse gases. In summary, simple thermodynamic calculations show that without greenhouse gases no living creature would survive on the Earth's surface, which would be completely frozen at an average temperature of approximately -20 °C.

It is evident that even a small change, say of a few percent, in the overall greenhouse gas concentration may induce a significant global climate change on the order of some degrees Celsius. The reduction of the greenhouse concentration by just 25% would plunge the Earth into a glacial era. Because of the great sensitivity of climate to changes in greenhouse gases, climate models need to correctly take into consideration the greenhouse effect to describe climate change because errors in this modeling can have significant consequences and lead to misleading conclusions. But how is this modeling now done?

As explained above climate models are today represented by large computer programs constructed to simulate the interactions among all elements believed to effect climate such as the atmosphere, oceans, land surface, vegetation and ice. These programs are then used for multiple purposes such as describing past climate as well as projecting future scenarios, and on a local scale they are used mainly for weather forecasts. There exist several kinds of climate models. The simplest ones are radiant heat transfer models, which are essentially energy balance models (EBMs). The most advanced and complex ones are coupled atmosphere-ocean-sea-ice-global climate models, which are usually called Global Climate Models or General Circulation Models (GCMs), and which contain many geophysical details.

From a mathematical point of view these models are complex networks of coupled differential equations that describe the physical and thermodynamical mechanisms of climate. Specifying their numerical solution require the specification of initial and boundary conditions, as well as imposing radiative forcing as energy input which are time dependent. The uncertainty in these conditions is dealt with by implementing the models multiple times with slightly different parameter values and studying the properties of the collection of simulation outcomes. Some model boundary conditions might consist, for example, in imposing the observed temperature of the surface of the ocean. Radiative forcing would include the total solar irradiance forcing and a set of forcing that correspond to an estimated radiative effect of changes in the greenhouse gas concentrations, volcano aerosol, changes of the albedo and other components. The estimate of the global average radiative forcing considered in the IPCC report [162] and their ranges from 1750 to 2005 is shown in Figure 1.19.

Before discussing the results catalogued in Figure 1.19 a reader can wonder why among the greenhouse forcing the one associated with water vapor is missing. This omission seems bizarre since the greenhouse effect of water vapor is significantly greater than that of all the other greenhouse gases combined!

The answer to this question is simple: modern climate models treat water vapor not as climate forcing but as climate *feedback*. In other words the models contain mathematical equations that allow the water vapor concentration in the atmosphere to automatically change as a response to the atmospheric temperature. Thus, for example, an increase in atmospheric temperature due to an increase of the radiative forcing leads to an increase in the water vapor content of the atmosphere through increased evaporation. The increased

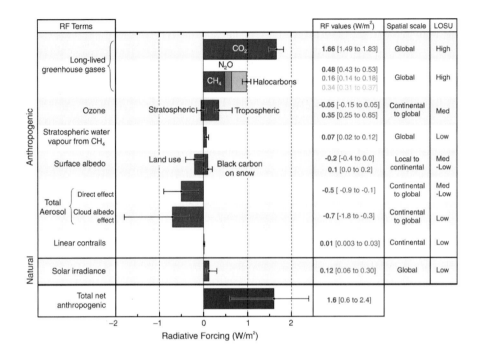

Figure 1.19: Global average radiative forcing estimates and ranges from 1750 to 2005 for anthropogenic carbon dioxide (CO_2), methane (CH_4), nitrous oxide (N_2O) and other important agents and mechanisms, together with the typical geographical extent (spatial scale) of the forcing and the assessed level of scientific understanding (LOSU). The net anthropogenic radiative forcing and its range are also shown. These require summing asymmetric uncertainty estimates from the component terms, and cannot be obtained by simple addition. Additional forcing factors not included here are considered to have a very low LOSU. Volcanic aerosols contribute an additional natural forcing but are not included in this figure due to their episodic nature. The range for linear constraints does not include other possible effects of insolation on cloudiness [from [162] with permission].

water vapor in turn leads to an increase in the greenhouse effect and thus to a further increase in temperature; the increase in temperature leads to still further increase in atmospheric water vapor; and the feedback cycle continues until an equilibrium is reached. When a feedback enhances the effect of some other climate forcing it is said to be a *positive climate feedback*. Alternatively, if a change in the environment leads to a compensating process that mitigates the change it is said that it is a *negative feedback mechanism*.

Later we develop the feedback concept further. For the time being we briefly discuss Figure 1.19 and attempt to understand the topological structure underlying the IPCC interpretation of global warming. First, we note that the forcing in the figure are divided into two major groups labeled *natural forcing* and *anthropogenic forcing*, which are calculated for the period from 1750 to 2005. On average the net anthropogenic component is estimated to be approximately 13 times larger than the natural one. Under the label *natural forcing* we find only the total solar irradiance forcing, that is the estimated forcing induced by the changes of solar irradiance 1750 to 2005. Under the label *anthropogenic forcing* we find the estimated radiative forcing associated with the changes in concentrations of CO_2, CH_4, N_2O and Halocarbons which are positive and, therefore, have positive contribution to global warming. The radiative forcing of the Ozone is divided into tropospheric and stratospheric with the former having a warming effect and the latter a cooling effect. The changes of surface albedo due to land use and to black carbon on snow; the land use for agricultural purpose have evidently increased over the last few centuries and the deforestation has had a cooling effect because the forests are darker than the ground and absorb more light, so with less forests more light is reflected back into the space; dirty snow instead has a warming effect because it absorbs more light, thus the terrestrial albedo is reduced. The increase of aerosols concentration, instead, has had a cooling effect; these values are all very uncertain.

The most salient features of Figure 1.19, as Figure 1.20 schematically reproduces, is the fact that the all radiative forcing, except for the solar one, are labeled as anthropogenic. The expectation is that without humans such forcing should be set to zero. Herein we do not discuss whether the intensity of the forcing used in the models is correct: people might disagree over whether these values should be larger or smaller. We merely notice that the above radiative forcing are a specific function of the variation in the atmospheric chemical concentrations. Indeed, a problem arises with the *anthropogenic* labeling of these forcing because they are deduced from the

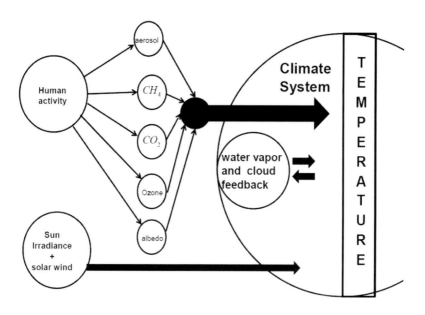

Figure 1.20: The topological structure underlying the IPCC interpretation of global warming implicit in Figure 1.19. The climate system is forced by a set of independent radiative mechanisms, which are divided into two groups: the natural forcing which consists of the total solar irradiance forcing; and the anthropogenic forcing that acts through five different independent channels (CO_2, CH_4, Ozone, Albedo and Aerosol which act in concert). Water vapor and clouds are modeled as climate feedback as indicated by the double-directional arrows. Global temperature is the outcome of the model. (N_2O and Halocarbons are omitted in the figure for graphical simplicity; the volcano effect is omitted because it is omitted in Figure 1.19).

observed change in the atmospheric chemical concentrations, and not from the estimated human emissions of the chemical elements. Note that the reason the counting starts in 1750 is because this year is arbitrarily identified as the beginning of the industrial era. The implicit assumption is that whatever chemical change occurred in the atmosphere since 1750 has to be ascribed to human industrial activity!

In other words, labeling as *anthropogenic* the climate forcing associated with measured changes of atmospheric greenhouse gases, changes in albedo and aerosols would imply that since 1750 the climate network itself cannot naturally alter such values in any substantial way. Above we have already observed that climate models implement the greenhouse effect of the water vapor and clouds by means of feedback mechanisms, that is, dynamically. The real climate network might contain CO_2, CH_4, ozone, albedo and aerosol feedback mechanisms as well, although the climate models presently being used do not contain them because they are still too difficult to engineer and for simplicity treat these components are modeled as purely external forcing. In a scenario where all greenhouse gases and other climate components are included dynamically, the climate network would not correspond to the model shown in Figure 1.20, but to that shown in Figure 1.21, where the newly added set of feedback connections among the elements drastically changes its topological structure. The latter topological model allows that not only human activity but also changes in solar activity and the climate network itself might alter CO_2, CH_4, ozone, albedo and aerosol chemical concentrations.

The greenhouse gas and additional feedback mechanisms (some of which might have a positive feedback effect, while others may have a negative one) are still very uncertain and not efficiently modeled, and therefore, are still absent in most global climate models. However, there is compelling evidence that these climate mechanisms really exist and do alter climate. To establish their existence requires temperature data from which the human factor is completely absent.

Climate and atmospheric composition history regarding epochs anterior to human civilization (that is, during periods when humans could not alter climate in any way) has been documented by analyzing the chemical composition of ice cores. An ice core is a cylindrical sample of the ice extracted from glaciers, such as from those in Antarctica and Greenland. Here the temperature is always below 0 °C and the snow never melts. So, over the years the snow and ice accumulate and recrystallize and trap air bubbles

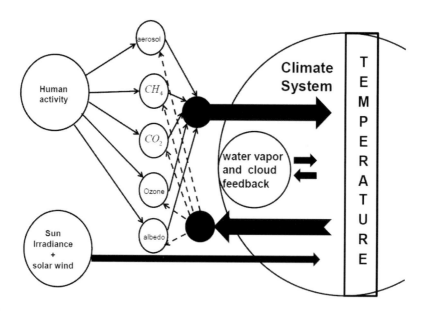

Figure 1.21: This sketch depicts the topological model of the terrestrial climate that includes additional feedback mechanisms over those shown in Figure 1.20. The climate network is forced by changes in solar activity, which is not limited to the Total Solar Irradiance alone but might include the effect of solar wind and UV on climate as well, plus the effect of anthropogenic activities and emissions that alter climatic components such as the Albedo and CO_2, CH_4, Ozone and Aerosols atmospheric concentrations. However, the climate system also contain several feedback mechanisms which are not limited to just regulating the feedbacks of water vapor and clouds, but in principle might effect any other climate component, as indicated by the additional arrows that from the climate network move within the climate components.

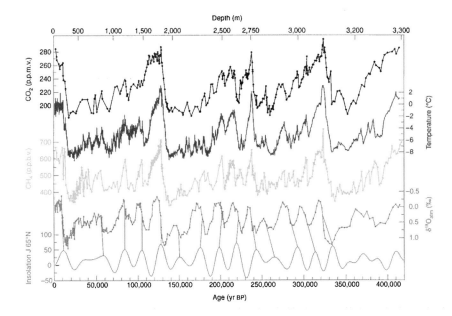

Figure 1.22: Climate and atmospheric composition history of the past 420,000 years from the Vostok ice core, Antarctica [126]. CO_2; Isotopic temperature of the atmosphere; CH_4; $\delta^{18}O_{atm}$; and mid-June insolation at 65° N (in W/m^2). The figure shows that the climate system contains strong CO_2 and CH_4 feedback mechanism, and that large climate variations are coupled to insolation variation.

from previous time periods. Thus, the depth of the ice core can be related to a specific epoch. It is possible to determine the atmospheric concentration of several greenhouse gases by analyzing the chemical composition of the air bubbles trapped in this ice. The concentrations of CO_2 and CH_4, aerosol, and the presence of hydrogen and oxygen isotopes provides a picture of the climate at the time.

The Vostok ice core in Antarctica is very deep and has been used to reconstruct an uninterrupted and detailed climate record extending over hundreds of thousands of years [126]. Some of these data are shown in Figure 1.22. The records indicate that the temperature reconstruction (of Antarctica) underwent large cyclic oscillations with a period of about 100,000 years and with a peak-to-through amplitude of about 10 °C. Thus, the Earth underwent long periods of glacial epochs that repeat almost periodically; being interrupted by short warming periods (less than 10% of the total). Human civilization has developed during the last warming period, that is, the last 10,000 years. The CO_2 and CH_4 concentrations were not constant during this time either, but underwent almost identical long-time oscillations. Analyzing the difference between the cold and warm epochs the CO_2 concentration could increase by approximately 50% and the CH_4 concentration could approximately double. These large oscillations have been associated with the change of insolation associated with the almost periodic variations in Earth's orbit known as the Milankovitch cycles [110]: see Figure 1.23.

The Vostok ice core data establish that the climate network contains strong greenhouse gas feedback mechanisms. In fact, CO_2 and CH_4 records follow the temperature record; a change in temperature induces a change in the CO_2 and CH_4 concentration that then reinforce the temperature change. Thus, these greenhouse gas feedback mechanisms are *positive*. Evidently these changes were not caused by human activity. In fact, the comparison between the Vostok ice core data and the Milankovitch cycles does suggest that variations of insolation (in this case due to terrestrial orbital changes such as its eccentricity) can cause significant climate changes. Because the Milankovitch cycles are predictable, we expect that the Earth will precipitate into another ice age in a few millennia. In any case, what concerns us here is that the climate contains strong natural greenhouse gas feedback that can be stimulated by changes in insolation. This finding reinforces our conviction that the climate network has a topological structure closer to the one shown in Figure 1.21 than that shown in Figure 1.20.

The argument that since 1750 these natural greenhouse feedback mechanisms have simply *disappeared* is not physical. Some of the natural mechanisms regulating the above greenhouse feedback are extremely slow and can play a role on geologic time scales only: for example, CO_2 removal via silicate weathering. However, other mechanisms are significantly faster: for example, a sea surface warming will increase atmospheric CO_2 through degassing [30]. Biological mechanisms can play a major role, but are still poorly modeled. The latter mechanisms refer to the balance between CO_2 uptake by land plants through photosynthesis and soil respiration production, and to the huge organic and anorganic marine CO_2 pump, which is a physico-chemical process that transports carbon from the ocean's surface to its interior [106]. The CO_2 pump occurs largely in the sunlit surface and may be easily effected

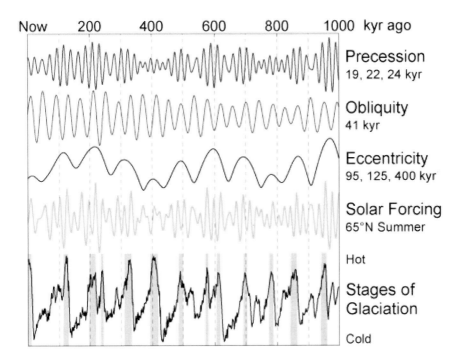

Figure 1.23: Milankovitch cycles and global climate. According to Milankovitch, the precession of the equinoxes, the variations in the tilt of the Earth's axis (obliquity) and changes in the eccentricity of the Earth's orbit are responsible for causing the observed 100,000 years cycle in ice ages. This happens because orbital changes modulate the amount of sunlight received by the Earth at different times and locations. The changes of the Earth's orbit can be calculated taking in account the gravitational interactions between the Earth, the Moon, and the other planets of the solar system. The data are from Quinn et al. [129] and from Lisiecki and Raymo [90].

by changes in temperature, cloud cover, ocean currents, nutrients availability, and both visible and ultraviolet radiation.

Climate can also be influenced by volcanic eruptions such as El Chichón (1982) and Mount Pinatubo (1991) eruptions that have been widely studied. Powerful volcano eruptions inject significant quantities of aerosols and dust into the stratosphere. Sulfur dioxide (SO_2) oxidized in the atmosphere to produce sulfuric acid droplets. These aerosol droplets gradually spread throughout the stratosphere over the year following the eruption, after which time their presence gradually decays. Stratospheric aerosols screen the sunlight and, consequently, the amount of solar radiation reaching the Earth's surface is reduced. Thus, a volcanic eruption causes a short-time cooling of the surface while the temperature in the stratosphere rises due to absorption of the solar radiation by the aerosols.

The intensity and the frequency of volcanic eruptions determine the impact that these geological events have on climate. Thus, climate models are forced also with an estimated radiative forcing associated with volcanic eruptions. The intensity of the radiative effect of these events is debated because all the mechanisms involved are still not well understood. Volcanic eruptions have an episodic nature, so their effect over long time scales can be neglected (this is why the table in Figure 1.19 omits them). However, an extremely large volcanic eruption might disrupt climate over a significantly long time (in a phenomenon known as *Volcanic Winter*) and materially effect the ecosystem.

There is a final issue regarding the anthropogenic effect on climate as implicit in the table shown in Figure 1.19 published by the IPCC. This issue is indeed a paradoxical one. In fact, in Figure 1.19 human activity is claimed to have been entirely responsible for the altered climate through the observed changes of CO_2, CH_4, Ozone, Albedo and Aerosols. The same models are then used to predict climate changes under several human emission scenarios for the 21st century as shown in Figure 1.17. The predicted consequences are that humankind might severely damage itself by causing a major climate change. Thus, the entire logic underlying the global warming debate is that not only can humans effect climate but that climate can effect humans as well.

What is paradoxical in the above IPCC logic is that such an argument is missing in the climate models used to forecast the anthropogenic effect on climate. In fact, the above logic implies that there exist climate feedback mechanisms that influence human activity that then alters climate. Indeed, the human network is part of the climate network and the effect of a climate change on the human network should be carefully investigated and included in the models. The net feedback of climate on humans may be positive or negative, and is undoubtedly temperature dependent as well.

The climate-human connection may have usually worked as a positive climate feedback [137]. Typically, a temperature increase would benefit humankind whose agricultural activity and emission would then increase and this would accelerate the warming. On the contrary, a decrease of solar activity may start a climate cooling causing periods of severe drought that may cause some civilizations to suffer famine and plague and ultimately to collapse with the abandonment of agricultural activity. This collapse would favor a reforestation that would yield to a higher absorbtion of CO_2 from the atmosphere and a further cooling of the climate.

The above mechanisms are historically confirmed by the fact that human civilizations have prospered during warm periods, and declined during cold periods. However, a too high temperature may be disadvantageous and detrimental for humankind (this is the logic behind the fear in the global warming debate). But if a too high temperature is detrimental for humankind, humankind will decline and human activity and emission would decrease and the climate would cool. Thus, above a certain temperature the climate-human connection would work as a negative climate feedback. Eventually, the anthropogenic feedback might even help to stabilize climate around an optimal temperature value. Neglecting these feedback mechanisms may lead the climate models to very misleading predictions. Human network response to global climate change may be the least understood feedback mechanism, but it may also be one of the most important to really understand climate change and its consequences for humankind!

A realistic topological model of the terrestrial climate may look like the more complex network shown in Figure 1.24. The Sun is the only strictly external forcing of the Earth's temperature: evidently a terrestrial climate change cannot alter solar dynamics. Volcanic activity is nearly an external forcing, but there may exist some coupling with solar activity, moon orbits and the climate network. The human network effects climate, but the climate, in turn, effects the human network. The albedo, CO_2, CH_4, ozone and aerosols can be altered because of human activity and also because of the solar forcing because they are partially feedbacks. Water vapor and clouds also provide climate feedback.

It is evident that the network shown in Figure 1.24 is quite different from that in Figure 1.20, which reproduces the topological model underlying the

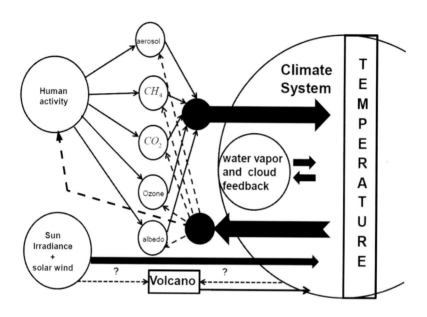

Figure 1.24: A realistic topological model of the terrestrial climate. The climate network is forced by changes in solar activity, which is the only sure physical external forcing. Volcano activity can be in first approximation considered an external climate forcing, but volcano activity might be partially triggered by solar activity and other climate mechanisms which is still a mystery. The human network forces climate through several mechanisms that alter climate components, but climate feedback mechanisms also affect thee components in several ways. Climate change affects the human network as well.

IPCC report implicit in the table summarizing the global climate change attributions (natural vs. anthropogenic) shown in Figure 1.19. But, what is the logical implication of the topological differences between the two models?

As we have explained the IPCC global climate change attribution estimates may be erroneous for quantitative reasons. For example, the water vapor feedback may be poorly modeled. If it is underestimated by existing models, new models with improved water vapor feedback would emphasized the solar effect on climate and would conclude that the net anthropogenic effect is currently overestimated. However, even assuming that the numbers in Figure 1.19 are correct, the fact that the topology of climate is represented by a network such as that shown in Figure 1.24 and not by the network shown in Figure 1.20 has significant consequences. A change in topology implies a change in how the several climate components are coupled and connected, and this greatly changes the way global climate change attributions (natural vs. anthropogenic) should be calculated.

Thus, the IPCC global warming (from 1750 to 2005) attribution according to which the anthropogenic component has been on average approximately 13 times larger than the natural one looks suspicious because it is based on a non physical topological model of climate dynamics; a topological model (Figure 1.20) in which climate itself cannot alter CO_2, CH_4, ozone, albedo and aerosols components by means of several natural feedback mechanisms and human activity itself is independent of climate. According to such a topology, the net anthropogenic contribution to global warming from 1750 to 2005 is just the sum of the independent contributions derived from a measure of the change of the above elements in the atmosphere, and the net natural contribution is given by direct total solar irradiance forcing alone.

On the contrary, according to the topology shown in Figure 1.24, the effect of the natural forcing on climate is likely to be larger than the IPCC attribution. This increase results from an additional positive solar-indirect radiative component. Solar changes alter the greenhouse gases via additional climate feedbacks whose net effect is likely to be positive because the greenhouse gases themselves have a net positive feedback. The solar forcing also enhances human activity by means of human-climate coupling mechanisms. The latter effects, although anthropogenic in nature, should be interpreted as a special kind of climate feedback to solar change. Authentic anthropogenic forcing should be the one that results from human activity that occurs despite the change of natural climate forcing, not the one that occurs because of it.

The above net positive indirect-solar radiative forcing should be evidently subtracted from the IPCC net anthropogenic attributed radiative forcing to climate change. In addition, the IPCC net anthropogenic attributed radiative forcing should be reduced by another positive net component because of the mutual feedback couplings among the greenhouse gases, which in Figure 1.19 were considered independent quantities.

Thus, we conclude by reasoning according to a pure topological perspective that the natural vs. anthropogenic global climate change attribution suggested by the IPCC report underestimates the solar-natural component and overestimates the anthropogenic one. The problem here is the incompleteness of the computational climate models due to the extreme difficulty of correctly including and modeling all relevant mechanisms. The reductionistic philosophical methodology implicit in the climate model approach, suggests a need to investigate the global warming phenomenon using an alternate method.

Chapter 2

Data

In this chapter we demonstrate that the design of scientific experiments requires a kind of thinking that is not so different from what we do every day, specifically thinking in terms of quantities. For example, in the weekly excursion to the Mall, imposed on the male by the female of the species, we engage in comparison shopping. In the comparison shopping of packaged goods the total price of a purchase is converted to the price per ounce to determine if a 24 ounce box of cereal costing \$4.80 is a better buy than a 19 ounce box costing \$3.95. Most middle class people do this kind of activity in buying anything from a pair of socks to a new house.

On a more abstract level one wonders whether a career decision should be made for maximum income or maximum satisfaction, or neither, or both. A facility with this kind of reasoning is a scientist's stock in trade, reflected in a willingness to put all problems into a quantitative context and in so doing cast them in a form that lends itself to mathematical analysis. By 'quantitative context' we mean designing an experiment to measure a specific quantity, whether it is something as direct as testing the physical fitness of a colleague, where objective measures exists, or as complicated as determining whether we can trust the promises of a political candidate, where no such quantitative measures exist. A well-designed experiment provides data to answer specific questions.

How do we obtain data, transform data into information and ultimately create knowledge from information? We would like to involve the reader in the process, so we hope to reveal the usefulness of the strategy of quantitative thinking. To those readers who think they can escape quantity in their lives we point to tempo in music, metric in poetry, balance in painting and

numbers on the stock exchange, as examples of the ubiquitous character
of quantity. Nobody can avoid quantity, but quantity shouldn't dominate
thinking either. The way to get beyond data is to seek out and encourage
regularity, while at the same time not dismissing the importance of variability.
The discipline of statistics was developed for just this purpose: to determine
how the erratic results of past experiments can be used to determine future
activities.

Experiment has historically been the backbone of science. It is the way
a researcher probes a biological, physical or social network to determine its
properties. In short, the procedure is to introduce the network into a laboratory and hold the laboratory environment fixed. The network of interest
could be: a piece of metal which appears to be stronger than other sub-
stances of similar weight; a chemical compound that thickens the blood; a
person with traumatic brain injury (TBI). The list can be carried to exhaus-
tion. By isolating the network in the laboratory the researcher can control
how s/he interacts with it during the experiment. S/he can: vibrate the
piece of metal and determine how long it takes for it to fracture; vary the
concentration of the compound added to the blood and note how the blood's
viscosity changes; record how the blood flow to the brain for TBI is different
from that of a normal healthy person. In each case a scientist measures a
quantity believed to answer the question s/he has asked with the experiment.

Most phenomena in the world do not lend themselves to laboratory experiment , so getting a positive result in a laboratory experiment does not
guarantee that the same result will be obtained in the uncontrolled real world.
This is called field testing; taking the experiment out of the laboratory and
determining if it will work in the field. Suppose the piece of metal in our
hypothetical experiment is found to be stronger than steel and lighter than
aluminium, under laboratory conditions. The metallurgist working on the
project is deliriously happy and in his glee proposes an application for his
new material, call it Rearden Metal. S/he proposes using Rearden Metal to
build the frame for a portable bridge and the task is taken on by the Army
Corps of Engineers. The Rearden Metal bridge is built in the desert and
the new properties of the material are taken into account in its design. The
first vehicle to cross the bridge causes it to collapse, putting the driver in
the hospital. It turns out that the strength of Rearden Metal is sapped by
prolonged exposure to the desert sun, causing it to nearly fold up under its
own weight. Of course one could adopt the view that the bridge was another
experiment designed to test Rearden Metal under actual working conditions.

We are sure that this view is what the metallurgist argued to the board of inquiry. However, this simple example shows the limit of the lab reductionistic approach according to which a physical network is assumed to conserve all its relevant properties once it is isolated from the larger network to which it is connected in the physical world.

In fact, most of what scientists want to understand about the world is outside the laboratory and not subject to the control of the researcher. A physical oceanographer may be fascinated by water waves, some of which can be generated in a water tank, but to really understand them s/he must go out on the ocean and experience the waves generated by the wind far from land. A meteorologist wanting to understand the essential nature of a hurricane cannot do that in the laboratory, but must measure the wind and rain as the storm tears through the countryside. The notion that the investigator can control the environment of the observation is absent from field research, so consequently the variability in the measurements is very much larger than analogous laboratory measurements. This variability in data is seen in both the globally averaged temperature of the Earth's surface and in the total solar irradiation reaching the Earth from the Sun. We argue subsequently that this variability and the unreliability of some data sets lead to the present impasse we have with regard to properly interpreting global warming.

Regardless of whether an experiment is being done in a laboratory or an observation is being made in the unconstrained world, measurements of the appropriate variable are what the researcher relies on to interpret what has occurred. The results of measurement are the data and that in our view forms the first of the three tiers of science. The purest may argue that there is no data without a theory of measurement, so that theory ought to be the first of the three tiers, but we do not think so. For the purposes of this chapter we consider data to be just what they are, that is, sensory input without benefit of interpretation or context. A loud sound can be the overture of Beethoven's *Fifth Symphony* or it can be the slamming of a car door, when it wakes the baby it does not matter; we still get up to sooth the little darling because of the loud noise.

We discuss data from the perspective of the controlled experiment in physics and discuss how this point of view has been applied far and wide outside the physical sciences in the eighteenth and nineteenth centuries, often with remarkable results. The interpretation of data was based on the measured phenomenon being a linear additive dynamic process, a condition that was often not met outside the physical sciences. In linear phenomena the response is proportional to the excitation; double the stimulus and the response is also doubled. However, two heads are not always better than one. The properties of a complex adaptive network determines the time series capturing the dynamics of a network variable. The variability of time series can reveal the dynamics of the network and so we discuss how to extract the underlying structure using the familiar ideas from statistics, such as averages and correlations.

Later in the chapter we concentrate on where these traditional measures breakdown, due to the complexity of the networks being studied. This complexity is manifest in scaling properties indicating a long-time correlation in the data, that is, a dependence of what is being measured on the history of the network. A number of physiological time series are used as exemplars of phenomena that have such history-dependent time series. In fact we argue that all physiological time series have this form because of the generic nature of complex networks in physiology.

Finally, we present data on the average global temperature, solar flares and sunspots. These data are seen to be inter-dependent using simple statistical measures. We discuss some of the difficulties associated with obtaining reliable data sets on a global scale, as well as the problems with using such time series to draw conclusions. One comparison that is quite remarkable is that between the statistics of the average global temperature anomalies and the fluctuations in the total solar irradiance (TSI),[1] as determined using a surrogate measure, the solar flare time series. It appears that the Earth's average temperature fluctuations inherit the statistics of the fluctuations in solar activity.

2.1   Physics as a scientific paradigm

A Greek philosopher observed that he could not enter the same river twice: evidently water flows. Here we observe that scientists cannot perform the same experiment twice. Significant differences always occur between the outcome of experiment/observation and the formulated expectation based on the prediction of natural law. No matter how carefully one prepares the

---

[1]The total solar irradiance is the energy the Sun emits per second on one square meter at a distance of an astronomical unit, which is the average distance of the Earth from the Sun.

initial state of an experiment, or how meticulously the variables of the experiment are controlled, the outcome always varies from experiment to experiment, leaving the scientist in the position of explaining why the result is not just the single number predicted. Natural laws purport to determine that there should be a unique outcome resulting from a given experiment starting from a given initial state, so when more than one outcome was observed from a sequence of apparently identically prepared experiments, the notion of error was introduced into eighteenth century science. The pejorative term *error* signifies the interpretation of the deviations of the experimental results from the predicted result. This connection between data (experimental results) and theory (predictions) was not always appreciated.

Experiments were introduced to suppress the natural variation in observations. The more stringent the control the less the variability in measurement, resulting in the notion that a network can be isolated from the environment and experiments could yield results independent of the environment. This turned out not be the case. While it was true that the influence of the environment on the experimental network could be vastly reduced, the network could never be completely isolated. Consequently, the effect of the uncontrolled environment is manifest in every measurement as error. This was especially true in the social sciences where the most important variables were often not even identified, much less controlled.

At a time when travel required horse-drawn carriages, reading involved candlelight or the fireplace, and gentlemen drank, gambled and whored, a new philosophy of the world was taking shape. In natural philosophy, which today we call the physical sciences, the clockwork celestial mechanics of Newton became the paradigm of science. Consequently, the predictability of phenomena became the hallmark of science and anyone seeking to be thought a scientist or natural philosopher emphasized the predictability of phenomena. A contemporary of the time, John Herschel, wrote in the Edinburgh Review in 1850:

> Men began to hear with surprise, not unmingled with some vague hope of ultimate benefit, that not only births, deaths, and marriages, but the decisions of tribunals, the result of popular elections, the influence of punishments in checking crime — the comparative value of medical remedies, and different modes of treatment of diseases — the probable limits of error in numerical results in every department

> of physical inquiry — the detection of causes physical, social, and moral — nay, even the weight of evidence, and the validity of logical argument — might come to be surveyed with that lynx-eyed scrutiny of a dispassionate analysis, which, if not at once leading to the discovery of positive truth, would at least secure the detection and proscription of many mischievous and besetting fallacies.

What Herschel was addressing was the new statistical way of thinking about social phenomena that was being championed by social scientists such as Quetelet, among others. As Cohen [26] pointed out, one way of gauging whether the new statistical analysis of society was sufficiently profound to be considered revolutionary was to determine the intensity of the opposition to this new way of thinking. One of the best-known opponents to statistical reasoning was the philosopher John Stuart Mill [111]:

> It would indeed require strong evidence to persuade any rational person that by a system of operations upon numbers, our ignorance can be coined into science.

In other words Mill did not believe that it was possible to construct a mathematics that would take the uncertainty that prevails in the counting of social events and transform that uncertainty into the certainty of prediction required by science. Of course there are many that today still share Mill's skepticism and view statistics as an inferior way of knowing, completely overlooking the fact that this is the only way we gather information about the world.

The scientists of the eighteenth and nineteenth centuries would claim that science ought to be able to predict the exact outcome of experiment, given that the experiment is sufficiently well specified and the experimenter is sufficiently talented. However, that is not what was and is observed. No matter how careful the experimenter, no matter how sophisticated the experimental equipment, each time an experiment was conducted a different result was obtained. The experimental results were not necessarily wildly different from one another, but different enough to question the wisdom of claiming exact predictability. This variability in the outcome of ostensibly identical experiments was an embarrassment and required some careful thought, since the eighteenth century was the period when the quantitative measurement of phenomena first became the same as knowledge. Consequently, if precision is knowledge, then variability must be ignorance or error; and that is what was believed.

It should be noted here that we have introduced an intellectual bias. This bias is the belief that there is a special kind of validity associated with being able to characterize a phenomenon with a number or set of numbers [155]. This particular bias is one that is shared by most physical scientists, such as physicists or chemists. I point this out to alert the reader to the fact that numbers are representations of facts and are not facts in themselves, and it is the underlying facts that are important, not necessarily the numbers. However, it is also true that it is usually easier to logically manipulate the numbers than to perform equivalent operations on the underlying facts, demonstrating why mathematical models are so important. Mathematical models of physical and social phenomena replace the often torturous arguments made by lawyers and philosophers concerning the facts of the world, with the deductive manipulations of mathematics applied to the representations of those facts. Consequently, a scientist can proceed from an initial configuration and predict with confidence the final configuration of a network, because the mathematical reasoning is error-free, even if the physical realization of the phenomenon is not.

The central fact regarding physical experiments was the variability in outcome and Carl Friedrich Gauss (1777-1855), at the tender age of nineteen, recognized that the average value was the best representation of the ensemble of outcomes of any given set of experiments. The general perspective that arose in subsequent decades is that any particular measurement has little or no meaning in itself: only the collection of measurements, the ensemble, has a physical interpretation and this meaning is manifest through the distribution function , which we take up in the next chapter. The distribution function, also called the probability density, associates a probability with the occurrence of an error in the neighborhood of a given magnitude. If we consider the relative number of times an error of a given magnitude occurs in a population of a given (large) size, that is, the frequency of occurrence, we obtain an estimate of the probability an error of this order will occur. Gauss was the first scientist to systematically investigate the properties of measurement errors and in so doing set the course of experimental science for the next two centuries.

Gauss formulated what came to be known as *The Law of Frequency of Errors* or just the law of errors, using the notion that the average or mean of a large number of experimental results is the best representation of an exper-

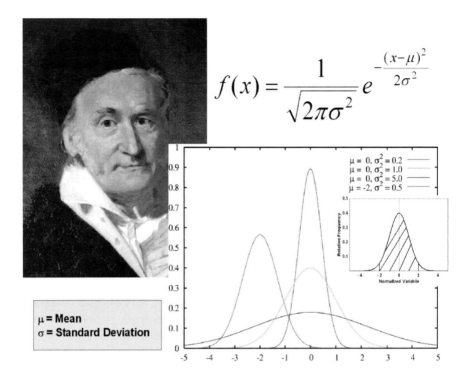

$$f(x) = \frac{1}{\sqrt{2\pi\sigma^2}} e^{-\frac{(x-\mu)^2}{2\sigma^2}}$$

Figure 2.1: Johann Carl Friedrich Gauss (1777-1855) is ranked as one of history's greatest and most influential mathematicians. He contributed significantly to many fields, including number theory, analysis, differential geometry, geodesy, electrostatics, astronomy, and optics. One of his numerous achievements was the development of the so-called *Gaussian distribution*, also known as the *Normal Distribution* or *Bell Curve*, for describing measurement errors. It is found that the normal distribution well approximates many psychological measurements and physical phenomena (like noise) very well. One of the reasons for its ubiquity is that according to the central limit theorem this distribution naturally emerges when an observation results from the sum of many small, independent effects. In addition, the normal distribution maximizes information entropy among all distributions with known mean and standard deviation, and for this reason it is the natural choice for the underlying distribution of data summarized in terms of sample mean and standard deviation.

iment [56] while the standard deviation[2] is a measure of the error magnitude. Gauss was particularly concerned with the effects of errors of observations of predicted future positions of celestial bodies such as the trajectory of Ceres [3], and in his analysis he laid the foundation for what we now know as statistics and the propagation of error. He also developed the method of least squares, a procedure used in all sciences to this day to minimize the impact of measurement error while fitting a set of observations with a given function that depends on one or more parameters.

Gauss determined that the errors from experiment, an error being defined as the deviation of an experimental observation from the average value of all such observations, form a bell-shaped curve centered on the average value, as shown in Figure 2.1. The bell-shaped curve is universal and quantified the notion that small errors occur more often than do large ones. Consequently the experimental value has the average value subtracted so the curve is centered on zero, and the new variable is divided by the standard deviation (width of the distribution) of the old variable and the new variable is therefore dimensionless. The cross-hatched region, between one standard deviation above and below the average value, contains 66% of the errors, while the interval between two standard deviations contains 95% of the errors (the deviations from the average value).

Thus, the curves in Figure 2.1 determine the relative amount of error of a given size in a collection of experimental data. Close to the average value, near zero in the figure, the curve has its maximum value indicating that a large fraction of the experiments yield results in this region. The central maximum not only implies that the average value is the most likely outcome of any given experiment, but also that the average value is not a guaranteed outcome of any given experiment. For large errors, far from the average value at the peak of the distribution, near the value of plus or minus two standard deviations, the height of the curve is nearly a factor of twenty smaller than

---

[2]The standard deviation is a statistical measure of the width of the bell curve. A very narrow distribution, small standard deviation, implies that the average is a good representation of the data, conversely a wide distribution, large standard deviation, implies the average is a poor representation of the data.

[3]The story about the prediction of the trajectory of Ceres includes a curious episode that highlights the prodigious mental abilities of Gauss. When he was asked how he could be able to do it Gauss replied, "I used logarithms." Then, the questioner asked him how he could have been able to look up so many numbers from the tables so quickly. Gauss responded: "Look them up? Who needs to look them up? I just calculate them in my head!"

it is at the center. This sharp reduction in the fraction of outcomes indicates that errors of this size are many times less frequent than are those near zero. Gauss developed this view of measurement because he understood that there ought to be a right answer to the questions asked by experiment. In other words, the outcome of an experiment, even though it has error, is not arbitrary, but rather the outcome follows a predetermined path described by the equations of motion established by Newton.

This argument neatly gathers the data from experiments, expresses data as information in the form of the distribution whose interpretation provides knowledge by quantifying how well the average value characterizes the experiment using the standard deviation. Gauss maintained that the physical observable, the quantity over which we actually have control in an experiment, is the average value of the variable and the fluctuations are errors, because they are unwanted and uncontrollable. It is the average value that we find in natural law and it was this attractive perspective that eventually lead to the application of the physics paradigm to the social and life sciences.

2.1.1   Psychophysics quantifies individuals

Attempts to understand complex social networks can often be traced back to the same investigators working to understand complex physical networks. In 1738, Daniel Bernoulli (1700-1782), the person who contributed so much to our understanding of hydrodynamics and the theory of probability, was interested in characterizing the social behavior of an individual and to do this he introduced a function intended to describe an individual's social well being: the Utility Function. Bernoulli reasoned that a change in some unspecified quantity x, denoted by Δx, has different meanings to different persons depending on how much x they already possess. For example, if x denotes a person's level of income, then the greater the level of income the less important is the magnitude of any particular change in income. Thus, if we indicate with $\Delta u = \Delta x / x$ the variation of utility accomplished, we have that the utility can be represented by a logarithmic function of the income x: $u(x) = c \log(x)$, where c is an opportune constant.

To make this argument more concrete, suppose two people invest money in the stock market. One person invests \$10 and another invests \$100. Then, it happens that both succeed in gaining \$10. It is evident that the former person has doubled his money while the latter has realized just a 10% profit. Who is happier? Both have made the same amount of money, but it is an aspect of human nature that we respond more strongly to the larger percentage change, which is to say, the greater $\Delta x/x$, the greater the response. Verifying this aspect of human nature occupied a number of scientists during the nineteenth and twentieth centuries, although it was taken as obvious by Bernoulli, that the former person would be happier over his 100% profit than the latter would be over his 10% profit.

Bernoulli posited that transactions made by individuals at different income levels are equivalent if the ratio is the same for both participants, that is, the transaction involves the same fraction or percentage of each person's income. The ratio is independent of the units used to measure x and therefore does not depend on the scale of the process. Thus, it does not matter if a person's income is in dollars, pounds or pesos; a 10% change is independent of the currency. So, utility is a measure of *relative* satisfaction. Given this measure, one may speak meaningfully of increasing or decreasing utility, and thereby explain some behavior in terms of attempts to increase one's utility.

In the nearly three hundred years since Bernoulli introduced the utility function science has learned that the social behavior of the individual cannot be represented by such a simple mathematical form. However, the properties of many biological and social networks can be so described, as we discuss subsequently. It is true that the simple mathematical form of the utility function cannot give a complete characterization of an individual's response, however it does seem to capture an essential aspect of that response, that being its scale-free nature. Everything we see, smell, taste and otherwise experience is in a continual state of flux. But the changes in the world are not experienced uniformly; our responses to the world's vagaries are not in direct proportion to those changes. In the last century the physiologist E. H. Weber, experimentally studied the sensations of sound and touch and determined that people do not respond to the absolute level of stimulation, but rather they respond to the percentage change in stimulation. Shortly thereafter the physicist Gustav Fechner resigned as physics professor because of an eye disorder contracted while studying the phenomena of color and vision, and after taking a Chair in Philosophy founded a new school of experimental psychology called *Psychophysics*. He determined the domain of validity of Weber's findings and renamed it the Weber-Fechner Law [49].

According to the Weber-Fechner Law if the change in variable is the stimulation R, then we respond to the relative change in the stimulation $\Delta R/R$ rather than to the absolute level of the stimulation ΔR itself. This evidently means that $S = c \log R$, where S is the sensation measured in some

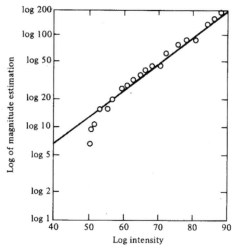

Figure 2.2: Gustav Theodor Fechner (1801-1887) was a pioneer in experimental psychology and the founder of psychophysics. He is credited to have created the formula $S = c \log R$ that proved the existence of a scientific connection between the body and the mind. The above equation was corrected by Stanley Smith Stevens (1906-1973) and the figure on the right shows psychological vs. physical intensity of sound magnitude estimates on log-log graph. The data are fit with a power law of the Stevens' type $S \propto R^c$, that on a log-log graph is a straight line with slope c. The measured scaling value of the power-law index is $c = 0.30$. [Taken from [136]].

unit and c is an opportune constant that must be separately determined by experiment for each particular order of sensibility. The Weber-Fechner Law can be expressed as follows:

In order that the intensity of a sensation may increase in arithmetical progression, the stimulus must increase in geometrical progression.

This early work supported the intuition of Bernoulli even though its basis was psychological/biological and not social. Scale-free functions seem to be the more natural way of characterizing psychological and social phenomena; however, Fechner did not get it quite right. In his famous paper entitled "To Honor Fechner and Repeal His Law," Stanley Smith Stevens (1906-1973) argued that the relation of sensation to intensity should be revised, that is, the logarithm should be replaced by a power law of the type $S \propto R^c$, see the discussion in Roberts [136]: an idea first put forward by Plateau in 1872. In any case, both Weber-Fechner's and Stevens' equations suggest the existence of a connection between the body and the mind that can be investigated using the methods of physical science making them susceptible to measurement and mathematical treatment. Thus, the importance of this discovery relies on the fact that psychology had the potential to become a quantified science contrary to the opinions of philosophers such as Immanuel Kant who had long stated otherwise.

An example of the application of the power-law response function is in terms of the subjective loudness of a sound (brightness of a light or other stimulus) relative to a reference sound. Suppose a reference sound is given an arbitrary value and a second sound, perceived as half as loud, is given half the value of the reference number. Similarly a third sound perceived as twice as loud is given twice the value of the reference number. The graph of these numbers for a variety of intensities versus the actual intensity of the sound is shown in Figure 2.2. The straight line, determined by a linear fit to the data, on this log-log graph paper clearly indicates a power law for the psychophysical function.

This curve fitting describes the regularity of the experimental data; it is not theory. In this case the estimate of the scaling index, determined by the slope of the straight line, is found to be $c = 0.30$. Other excitations that have been shown to satisfy similar power-law behavior include the intensities of: light, smell, taste, temperature, vibration, force on handgrip, vocal effort and electric shock among a myriad of others [136]. Each of these disparate phenomena has a positive power-law index given by the slope of the estimate intensity curve.

In his encyclopedia contribution to psychophysical scaling Roberts [136] points out that the psychophysical law involves new predictions as done by Stevens [166] in his use of cross-modality matching. Consider two sensor modes, both of which respond with a power-law sensitivity, and design an experiment such that rather than associating the intensity of a stimulus with an arbitrary reference number, the level of excitation is characterized through the second modality. For example, the brightness of a light can be adjusted to match the loudness of a tone; double the intensity of the sound and the subject doubles the brightness of the light. In this way the intensity level of the two modalities can be graphed versus one another on log-log graph paper to obtain a straight line whose slope is related in a known way to the power-law exponents of the two modalities. These have come to be called *equal sensation graphs*, such as shown in Figure 2.3.

The development of metrics was one of the key scientific discussions arising in formulating psychophysics, in particular, how to relate physical to psychophysical measurements. Such metrics determine the possible forms of the psychophysical laws that have either a ratio scale or an interval scale and the difference between the two kinds of scales can sometimes be subtle. An example of a ratio scale is mass, where we freely change the units of a mass and therefore characterize it as one thousand grams or as one kilogram, simply through a multiplication factor of one thousand. Temperature is also a ratio scale if absolute zero is included; however this is not the case for the most common units, Centigrade and Fahrenheit. In addition to scaling the units, 9/5 in going from °C to °F, there is a shift in the freezing point (32 degrees) to go from the freezing point of water in one to the other and consequently we have an interval scale. The psychophysical law summarizing the data reflects this difference in metrics resulting in the power-law function.

This focus on metrics is equally important today and will continue to be important in the future, in part, because the measure of a given CAN determines what can be known about it and the conclusions that can be drawn concerning its place in nature. The notion of a metric is bound up in the ideas of measurement, information and knowability. For example, health is a basic concept in medicine; consequently, a physician determines the health of an individual by measuring such quantities as heart rate, blood pressure and breathing rate. Notice that each of these measured quantities is an average value, in keeping with the Gauss paradigm, which implies that

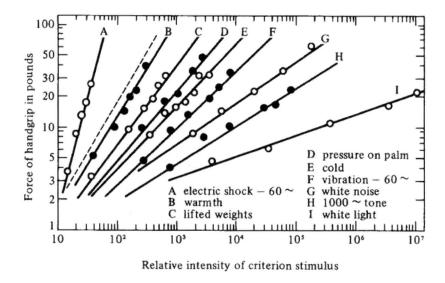

Figure 2.3: Equal sensation functions. Data obtained by matching the force of a handgrip to various criterion stimuli. Each point represents the median force exerted by ten or more observers to match the apparent intensity of criterion stimulus. The relative position of a function along the abscissa is arbitrary. The dashed line shows a slope of 1.0 in these log-log coordinates, which imply a power-law relation. [from [136] with permission].

physiology is the superposition of a number of linear additive, interactive networks. Research in the last two decades shows that physiology is, in fact, made up of a number of nonlinear multiplicative interactive networks, suggesting that for the past two centuries medicine has not been using the best measures of health and this turns out to be the case [187]. More appropriate measures of health are introduced subsequently and the need for additional research into a science of networks is discussed.

2.1.2   Sociophysics quantifies groups

In the preceding section some of Bernoulli's thoughts about modeling social phenomena using his utility function were mentioned. A somewhat more rigorous statement of what Bernoulli had in mind is that individuals make decisions based on maximizing their expected utility in a variety of contexts. People decide how to get the most 'bang for their buck'. The idea of expectation value is important because it assumes uncertainty; either uncertainty in the information used as a basis for a decision and/or uncertainty in the consequences given a particular decision. Uncertainty and randomness go together, the latter being the mathematical quantification of the former. Consequently, we compare what we think the value of two alternatives will be, using a personal utility function based on our experiences. For example, a student may select an elective in Art History, rather than Symbolic Logic, because he enjoys learning about art and artists and finds logic boring. Moreover, he knows that the art information will stay with him, whereas the logic formalism will evaporate when the course ends. On the other hand, the girl he is dating would make just the opposite choice. More formally, if the expected utility of choice A is greater than that of choice B, then it is prescribed that we select A over B using the utility function. Moreover, any variation of the expected utility vanishes and this places restrictions on the statistical properties of certain consequences.

Note that the expected utility argument revolves around knowing the utility function for the process of interest. Bernoulli's choice of utility function was the logarithm of the probability which associates a slow monotonic increase in the utility of the consequence with a decrease in the probability of the occurrence of the event. The lower the probability of an outcome occurring, the more useful it will be when it finally does occur. Different kinds of utility functions can be constructed. However, the logarithmic form is of interest because the expected utility function in this case becomes the entropy for the process, in which case, the expected utility requirement of Bernoulli becomes the principle of entropy maximization. The latter theory was developed in the twentieth century and we have occasion to discuss it in detail in our considerations on information.

Since the time of Francis Bacon (1561-1626) it was acknowledged that experimental science (empiricism) is based on inductive reasoning from observed fact to mathematical axiom to natural law. Note that scientific inquiry is justified by the *belief* that laws of nature do exist, and, therefore, attempting to discover them is a worthy activity. The problem faced by investigators was to establish a proper methodology for such an inquiry. The Baconian scientific method based on the above inductive methodology supplanted other methods such as those connected with the occult trends of hermeticism and alchemy. In so far as society could be understood in the same way as natural phenomena, it too must be similarly based on a given set of laws, even if these laws regulating society would remain a mystery.

It is, after all, the existence of these laws and the facility with them that enables scientists to make predictions and thereby obtain new knowledge. So, before the time of Newton (1643-1727) the notions of order in the world, determinism and prediction became inextricably intertwined so that regular patterns were interpreted as the natural expression of reality's lawful character. Some argue that the most important contribution of physics to the understanding of society was the idea of causality and that events, even those that are not subject to experiment, are lawful in their occurrence. Moreover, even when phenomena are not subject to controlled experiment, data regarding their nature may still be obtained.

The quantitative analysis of social data began with Graunt's *Nature and Political Observations Made Upon the Bills of Mortality* in 1667 [60]. Unlike the other observers of his time, who were most often men of means and therefore had the leisure time to do experiments and accumulate data, John Graunt (1620-1674) was a haberdasher of small wares. The leisure he had was the result of long hours, hard work and a dexterous mind. As pointed out by Montroll and Badger [115] his small pamphlet was the first quantitative examination of the affairs of man and contained the raw data from parish records of births and deaths along with the cause of death. Without going into the details of his statistical analysis of the data it is useful to record some of the conclusions drawn from his analysis [60]:

> that London, the metropolis of England, is perhaps a Head too big for the Body and possibly too strong: that this head grows three times as fast as the Body to which it belongs ... that old streets are unfit for the present frequency of Coaches... that the fighting men about London are able to make three great Armies as can be of use in this Island

Regarding his own work Graunt states:

> It may now be asked, to what purpose all this laborious buzzing and groping?.... To this I might answer in general by saying, that those, who cannot apprehend the reason of these Enquiries, are unfit to trouble themselves to ask them.

The inventor of *social physics* and one of the most influential of its nineteenth century proponents was Adolphe Quételet (1796-1874). During the period from 1820 to 1850 he did more than any other to develop the infrastructure for a science of man and society using the physics paradigm. His ambitious attempt to base the physical, social and intellectual aspects of humanity on the same set of laws ultimately proved unsuccessful, but not for lack of effort. In his book *Système Social* [128] Quételet analyzed vast amounts of data and found repeated statistical regularities such as the frequency distribution of military draftees' heights being well-fit by the Gaussian error distribution from astronomy. Similar observation induced him to proceed directly from analogy and to argue that the variability in the social arena are like the multiple, small, perturbing forces that produce fluctuations in celestial observations, so that the observations are randomly scattered about the true celestial position.

The end result of all Quételet's arguments was the concept of the *average man*, in which the characteristics of being human are described by statistical regularities in the moral domain such as crime, suicide and marriage and in the physical domain such as height, weight and strength. In all these areas the characteristic is described by an average quantity such as a rate and the variability of the characteristic by the Law of Frequency of Error, the Gaussian distribution. As observed by Oberschall [121]:

> The stability of social order is thus ultimately rooted in the invariant traits and disposition of the *Homme Moyen*.

The notion of the *average man* was ultimately discredited but not before it permeated the thinking of the nineteenth century and found its way into many of the popular concepts of the twentieth century. We shall take up these aberrations at the appropriate places and discuss how they have limited our understanding of society and the nature of social networks.

Another of the nineteenth century savants that fearlessly applied statistical reasoning to a variety of social phenomena was the English eccentric Sir Francis Galton (1822-1911). He was so enamored with the Law of Frequency of Error that he wrote:

> I know of scarcely anything so apt to impress the imagination as the wonderful form of cosmic order expressed by the *law of frequency of error*. The law would have been personified by the Greeks and

deified if they had known of it. It reigns with serenity and in complete self-effacement amidst the wildest confusion. The larger the mob, and the greater the apparent anarchy, the more perfect is its sway. It is the supreme law of unreason. Whenever a large sample of chaotic elements are taken in hand and marshaled in the order of their magnitude, an unsuspected and most beautiful form of regularity proves to have been latent all along.

In the above quotation Galton points out that the normal distribution has a regularity and stability that is characteristic of a network as a whole and is quite distinct from the variability observed in the individual network elements. This property is, of course, true of other limit distributions, as we shall see subsequently, when we discuss the properties of random networks. If limit distributions did not exist then the stable behavior of societies could not be expected, thermodynamics would be an illusion and indeed the printed word would drift up from the page in a world without ordering principles. At the other extreme, the erratic behavior of a single molecule or a single individual does little to influence the general behavior of a network. However, there are instances in which the single molecule like the single person can create a cascade of events which overwhelmingly changes a network's character, the initiation is referred to by the contemporary term 'tipping point', or as a phase transition in physics. We consider this complex situation subsequently, as well.

At the end of the nineteenth century the separation of phenomena into simple and complex was relatively straightforward. A simple network was one that could be described by one or a few variables, and whose equations of motion could be given, for example, by Newton's laws. In such networks the initial conditions are specified and the final state is calculated (predicted) by solving the equations of motion. The predicted behavior of the network, typically a particle trajectory, is then compared with the result of experiment and if the two agree, within a pre-established degree of accuracy, the conclusion is that the simple model provides a faithful description of the phenomenon. Thus, simple physical networks have simple descriptions. Note that we do not choose to bring up nonlinear dynamics and chaos here, since these concepts did not influence the nineteenth century scientists applying physics to the understanding of society. However, we find that the modern understanding of complex networks does require the use of these modern techniques.

CAPTAIN JOHN GRAUNT

Figure 2.4: Francis Bacon (1561-1626) (top-right), Sir Francis Galton (1822-1911) (top-left), John Graunt (1620-1674) (bottom-right) and Adolphe Quételet (1796-1874) (bottom left). The Baconian scientific method based on an inductive methodology and statistics have been applied to understand sociology. The notion of the *Homme Moyen* (the average man) has been introduced while individual variability has been interpreted as noise or error.

As more elements are added to a network there are more and more interactions, and the relative importance of any single interaction diminishes proportionately. There comes a point at which the properties of the network are no longer determined by the individual trajectories, but only by the averages over a large number of such trajectories. This is how the statistical picture of a phenomenon replaced the individual element description. The erratic single trajectory is replaced with a distribution function that describes an ensemble of trajectories and the equations of motion for individual elements are replaced with an equation of motion for the probability distribution function. Consider the image of a crowd following the streets in any major city at lunch hour. The individual may move a little to the left or to the right, s/he may walk a little faster or slower, but the ebb and flow of the crowd is determined by the average movement along the street which sweeps up the individual. Looking out the window of a twenty story building reveals only the flow and not the individual differences. This view from above is what the social physics of the nineteenth century intended to provide, with individual variability being interpreted as noise or error.

2.1.3   Econophysics quantifies exchange

The present generation of physicists marks the advent of econophysics with the work of Stanley and his group [164], but this application of physics to the understanding of world economics and financial markets actually began towards the end of the nineteenth century with the investigations of Vilfredo Pareto (Figure 1.6). He (Pareto) believed that even in sociology there exists an objective world, which can be determined by our five senses and only understood through the use of empirical data that is interpreted in terms of mathematics, as discussed in the next chapter. He was part of a nineteenth century tradition that saw society as being a complex network made up of a number of interdependent parts and these parts were as tangible to him as the separate parts that make up the human body are to a physician. In fact Pareto often used physiologic analogies in explaining the workings of society. In Pareto's view a change in any one part of a social network was communicated throughout the network by mutual interactions, sometimes resulting in small local changes and at other times resulting in global changes that modify the network as a whole. This multiple network view lead him to the development of the concept of social forces that, in direct analogy with physical forces, act to maintain social organizations and to insure slow, gradual changes from one state of society to another. His outlook on social stability glossed over revolution, so prevalent in nineteenth and twentieth century Europe, in favor of the homeostatic response of give and take to maintain social equilibrium. Internal forces arise to balance the externally generated perturbations agitating society and eventually subdue them. The state of a healthy society, like that of a healthy body, was a kind of dynamic-equilibrium.

Pareto's world view was reinforced by the structure he found when he examined sociological/economic data; structure that he believed to be universal, that is, patterns in society that arise in all societies at all times. The data he examined initially was the income level of populations from a variety of social states at a number of different times. According to a common but naive understanding of fairness based on the concept of equality, one might have expected that each of these societies could be characterized by a frequency distribution located around a given average income. Furthermore, this average income would be typical of the particular social order with some individuals earning a little more than average and some individuals earning a little less, resulting in a Gaussian distribution to represent how income was shared in the society, see Figure 2.5. In such a world differences between societies could then be gauged by comparing the average incomes and the widths in the distributions of income. However the world did not accommodate this picture and income turned out to be distributed in a surprisingly different way.

Pareto did not hypothesize about the underlying mechanisms producing suffering and the inequities in society. However, given his engineering/scientific training he did believe that society followed general laws made manifest within a political structure and that these laws could be quantified in the same way natural laws were codified in physics and chemistry. To verify his scientific view of society he collected data that guided his interpretation of what is important in society and determined what would be the form of these laws. Consequently, he was a scientific observer in that he gave preeminence to data and its interpretation and was very cautious in formulating theoretical explanations about what he observed. The tension between what he observed and the idealistic theoretical notions of what society should be like are still with us today.

Pareto discovered that the distribution of income was neither symmetric nor bell shaped, as it would be for a Gaussian distribution; instead the income distribution extends toward greater and greater values, as shown schemati-

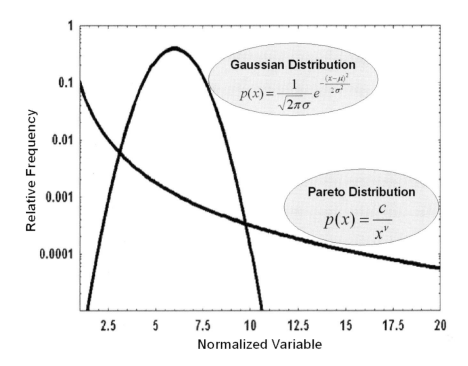

Figure 2.5: A Gauss bell-shaped probability density is compared with an inverse power law of the form empirically determined by Pareto to describe the distribution of income in western societies. Note that this is log-linear graph paper where the Gaussian curve appears as a parabola. On a log-log graph the Pareto curve would appear as a straight line with a negative slope.

cally in Figure 2.5. Indeed, Pareto could analyze income data referring only to the upper class that constitute a small percentage (less than 5%) of the total population which present an inverse power-law distribution. For the lower to middle class the distribution of income does present a bell shape, but it is a skewed one that can be fitted with a Gamma distribution: see Figure 2.6.

In any case from the comparison of the two distributions it is clear that the Gaussian distribution emphasizes those events that are near the center of the distribution, the mean value, and the Pareto distribution appears to completely disregard this central region and continue outward forever. This comparison is consistent with the everyday experience that we will never meet someone who is three times our height, but we can certainly meet someone who makes ten times, or a hundred times or even a million times our salary. Consequently the average income is not a useful measure of the empirical distribution because of this very long tail. This research first appeared in 1896 in Pareto's two-volume work [124] and continues to impact our understanding of society today, influencing even those who do not know his name.

The concept of dynamic-equilibrium is behind Pareto's theory of the circulation of elites according to which the history of society consists of a succession of elites whereby those with superior ability in the prevailing lower strata can, at any time challenge, and eventually overcome, the existing elite in the topmost stratum and replace them as the ruling minority. Nevertheless, although many replacements have occurred in human history the distribution of income for the upper class has always resembled an inverse power law shape where a small minority owns the great majority of wealth. The realization of the existence of a social/economic mechanism responsible for this phenomenon led Pareto to be very suspicious of the Communist promise that a revolution conducted by the largest class of proletarians would have successfully resulted in a society based on economic equality.

Thus, at the end of the nineteenth century there emerged the first of the empirical studies establishing the dangers of blindly applying to the social arena the distribution that had worked so well in physics. The distribution of Gauss was superseded by that of Pareto, in principle if not in practice. Perhaps, if the implications of Pareto's discovery were fully understood the twentieth century could have avoided some of the most atrocious and bloody attempts in human history to impose on some societies an ideological economic structure. Ironically this ideology claimed to be scientific, but that claim was based on an erroneous concept of political fairness supported by an

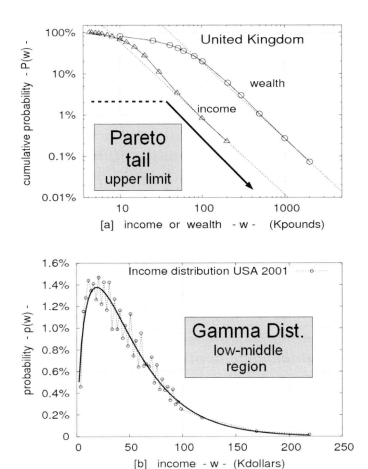

Figure 2.6: [a] Cumulative wealth (1996) and income (1998-1999) distributions in the United Kingdom [141]; the income of the upper class (less than 5% of the population) present a distribution that is well fit with a Pareto's inverse power law ($P(w) \propto w^{-\mu}$, where w is the level of income and $P(w)$ is the fraction of the population with that level of income) with an index of $\mu = 1.85$. [b] Income distribution (2001) in the USA, for the lower to middle classes (approximately 95% of the population) the distribution is well fit with a Gamma distribution ($p(w) = 0.48w^{0.57} \exp(-0.03w)$). (The abscises are in thousand pound and dollar units, respectively).

ideal of economic equality that was compatible with a symmetric bell-shaped distribution, for which there was no empirical evidence.

2.1.4   Biophysics quantifies life

In the nineteenth century biophysics, or physical biology, had been in the scientific shadows for a long time and as a discipline it has a number of candidate origins. Our personal favorite is Jan Ingen-Housz (1730-1799), who studied medicine, physics and chemistry in Leyden and subsequently became a Doctor of Medicine, but where and when he obtained his degree remains obscure. He practiced medicine in Breda, Vienna and London, being the first to introduce the vaccine against small pox into Austria by inoculating the Royal family. Probably his most important discovery was the explanation of the process of photosynthesis, the property of plants to use sunlight to facilitate the exchange of gas with the atmosphere. He determined that under the influence of light the green part of plants, such as the leaves, exhale oxygen and absorb carbon dioxide. The non-green parts of plants, the flowers and the fruits, almost continuously exhale carbon dioxide and absorb oxygen, as do the green parts in the absence of light. Photosynthesis and plant respiration is evidently one of the most important natural mechanisms for controlling the CO_2 and O_2 levels of concentration in the atmosphere. Evidently, photosynthesis is a very important and complex process in nature, however some of its phases are still not completely understood. Ingen-Housz treated the biological domain as being just another area of investigation for the methods of physics.

Random fluctuations in time series for natural phenomena are often refereed to as Brownian motion, after the Scotch botanist Robert Brown. It is not well known that Ingen-Housz was the original discoverer of Brownian motion. A full half century before Brown observed the erratic motion of pollen motes in a glass of water Ingen-Housz put powdered charcoal into a dish of alcohol and observed the same effect. It is only because someone told Einstein about Brown's experiments while he was writing the sequel to his 1905 paper on diffusion, and consequently he (Einstein) commented that the work of Brown might be related to the phenomenon of diffusion. It was in this way that the term Brownian motion was born; it could just as easily have been Ingen-Houszian motion, if Albert had been better read.

In his remarkable book *The Elements of Physical Biology* [97], Lotka defines biophysics as follows:

> In introducing the term Physical Biology the writer would suggest
> that the term Biophysics be employed (as hitherto) to denote that
> branch of science which treats of the physics of individual life pro-
> cesses, as exhibited in the individual organism (e.g., conduction of
> an impulse along nerve or muscle); and that the term Physical Bi-
> ology be reserved to denote the broader field of the application of
> physical principles in the study of life-bearing systems as a whole.
> Physical biology would, in this terminology, include biophysics as a
> subordinate province.

It appears that in 1924 there was still some debate over the proper domain of biophysics. Indeed, biophysics is an interdisciplinary science that
applies general theories and methods of physics to questions of biology. One
may argue that of the four areas: socio-, psycho-, econo- and bio- physics;
the last is where the ideas and techniques of physics have been the most
fruitful to date. This is probably due to the fact that biophysical phenomena
present an extended affinity with traditional physical and chemical phenomena. This seems evident if we list some of the most important biophysicists
and their discoveries: Luigi Galvani (1737-1798), who discovered bioelec-
tricity; Hermann von Helmholtz (1821-1894), who was the first to measure
the velocity of nerve impulses; studied hearing and vision; Sir Alan Lloyd
Hodgkin (1914-1998) and Sir Andrew Fielding Huxley (1917-living), who develop a mathematical theory of how ion fluxes produce nerve impulses; Georg
von Békésy (1899-1972), for his research on the human ear; Bernard Katz
(1911-2003), who discovered how synapses work; Hermann Joseph Muller
(1890-1967), who discovered that X-rays cause mutations; Linus Pauling
(1901-1994) and Robert Corey (1897-1971), who discovered the alpha helix
and beta sheet structures in proteins; Fritz-Albert Popp (1938-living), who is
acknowledged as a pioneer of biophotons work; John Desmond Bernal (1901-
1971), who is known for pioneering X-ray crystallography of plant viruses
and proteins; Rosalind Franklin (1920-1958), Maurice Wilkins (1916-2004),
James Dewey Watson (1928-living) and Francis Crick (1916-2004), who were
pioneers of DNA crystallography and co-discoverers of the genetic code; Max
Perutz (1914-2002) and John Kendrew (1917-1997), who were pioneers of
protein crystallography; Allan Cormack (1924-1998) and Godfrey Hounsfield
(1919-2004), who developed computer assisted tomography; Paul Lauterbur
(1929-2007) and Peter Mansfield (1933-), who developed magnetic resonance
imaging.

Thus, since the eighteenth century biophysics has had the intent to lift biology out of the realm of the descriptive into that of the quantitative following the path of physics. This was done by dissecting biological networks into parts and studying the physical properties of those parts. However, as science has progressed and the requirement to study biological networks in their entirety was imposed we may find that each of these attempts to directly apply the strategies of traditional physics outside its original domain were less successful than believed.

Indeed, social, economic and biological phenomena in their entirety may more properly be described in terms of complex adaptive networks. Here, the traditional physical methods are less satisfactory because of the intrinsic complexity of these networks and novel investigative strategies need to be developed.

2.2   Time series

As we explained in the previous chapter global warming is the hottest political/scientific debate in the world today. In its simplest form the argument states that we have a record of the average temperature at the Earth's surface for the last few hundred years. The temperature values are obtained using both direct and indirect measures. The point is that the average temperature at the Earth's surface is not constant in time, but has both erratic and regular changes from one year to the next, from one month to the next and from one day to the next. The arguments over what has caused these changes are long and detailed and the conclusion is that the regular component of these data shows a steady increase in the Earth's average temperature for the last 30 years; less than one degree Centigrade.

A certain segment of the scientific community maintains that this most recent temperature increase is due to the pollutants that human beings have released into the atmosphere over the past century. Another group of scientists maintain that natural phenomena contribute at least equally to this temperature increase. Paradoxically both positions use the same data, which are expressed as time series. But who is right? Evidently who is right cannot be determined using the data alone, but requires the information based on the data and subsequently the knowledge resulting from the information. But first let us try and understand what is meant by data and how that differs from information and knowledge.

It is clear which group has been doing the better public relations regarding global warming and its causes over the past decade; those favoring the fallibility of the human network. The second group has been quiet, in part, because scientists have historically shunned the public stage in favor of the less visible, but more substantial world of scientific publications. However, there comes a point where social pressure forces even the most reticent scholar to enter the public debate. But the second group is at a disadvantage because their arguments require an informed populace, not informed about the issue; but informed about the science. Since various governments are passing laws in response to the perceived threat of global warming, affecting people across the globe, it is reasonable to attempt to understand the basis of the supposed threat. Consequently, one must learn about time series that characterize the dynamics of complex phenomena, that is, about the form of the data upon which all these decisions are being made.

2.2.1 Measures and data

A time series is a set of measured values of a dynamic observable collected over time, usually at equally spaced intervals. The observable could be the Earth's globally averaged temperature, the total solar irradiance, the average CO_2 concentration in the Earth's atmosphere, the number of people with AIDS in Africa, the sizes and number of earthquakes in California, as well as many more; the only limitation is the imagination of the researcher. It is worth pointing out that the data discussed here are one step removed from thermometer readings, pressure indicators, and other instrumentation being discussed. Temperature recording stations are scattered over the globe and the Earth's surface temperature at a given time is defined by averaging over all these measurements at that time. This averaging is done from one time point to the next to define the time series for the Earth's surface temperature.

The important aspect of all these measurements is that they are taken over time. We interpret time series as being due to an underlying dynamical process; one that may be directly or indirectly related to the quantity being measured, a relationship that is not necessarily known at the time of the measurement. The theory leading to a dynamic interpretation that would enable researchers to make predictions concerning data not yet obtained constitutes what is generally regarded as knowledge. But here care must be taken because recognizing a simple pattern may enable a researcher to predict based on that pattern, often without real understanding. Understanding only comes when the pattern is explained by means of a general theory.

For example, in a simple physical network Newton's laws of motion provide the theoretical background to properly interpret the data. The skid marks left on the pavement can be analyzed to determine with certainty the speed of a car prior to an accident, whereas the past trajectory of a hurricane cannot be used to determine with certainty where the storm will strike next. Unlike the dynamics of the car, the dynamics of the hurricane is too complex for exact prediction. Consequently we have information about the storm in the form of probabilities, but we lack the certainty of knowledge with which to make definite predictions.

Measurements should be equally spaced in time whenever possible, using a reliable clock, and recording the value of an observable once every millisecond, or every day, or once every month, or whatever is appropriate for the network of interest. But it is not always possible to record with such regularity. Some processes change continuously in time such as the Earth's temperature; others are discontinuous and intermittent in time, jumping from one value to another without a smooth progression through intermediate values, such as the jagged fingering of lightening from clouds to the ground. The time series in the above two cases can be very different and may present very different properties. The way we sample the process ultimately determines the form of the time series.

Data regarding continuous phenomena can be obtained and recorded in a number of ways: 1) continuously, as with a strip chart recording a beating human heart recorded by electrocardiograms (ECGs) or the seismograph tracing monitoring earthquake activity, to name two analogue signals; 2) discretely, as in recording the time intervals between heart beat maxima or the time between quakes of magnitude above a given strength; to name two digital signals; 3) aggregating the data to obtain the total intervals or averages, the average earthquake activity per month or per year, the average heart rate per minute or per hour. A continuous data stream is often converted into a discrete data set, since a computer can only take data in discrete packets. This is what an analogue (A) to digital (D) converter does in the data collection and storage process; the A to D converter transforms a continuous, analogue signal into a discrete, digital signal. For example, a 100 Hz sampling rate means that every second the value of the continuous signal is recorded 100 times, with the interval between samples being equal to 0.01 seconds. Whether this sampling rate is too high or too low depends on the time variations of the phenomenon being investigated. The test is that typically data should contain multiple points from each feature of the underlying process, an erratic process with sharp spikes requires significantly more data points than a slowly varying process.

The form in which the data is collected and analyzed is selected to highlight different properties of the phenomenon being investigated. Therefore methods 2 and 3, rather than compromising relevant data, are chosen because they correspond to the form of the questions being asked. It is by such manipulation of the raw data that statistical measures are constructed and transformed into information about the phenomenon being probed.

Examples of discrete processes, those that do not change continuously in time but jump from one level to another, and therefore lend themselves to digital processing, are grade classification in schools (K through 12, elementary through high school), a person's age, and a person's yearly salary. In fact, most of the ways we organize our lives and classify our socialization involves quantization, that is, the compartmentalizing of processes and events into easily identified discrete elements. Learning and understanding are both continuous processes that are difficult to quantify, so we invented testing, grade levels and graduations to delimit this ongoing process into smaller and more manageable parts, so as to keep us from being overwhelmed by the process. This quantization of data is how we understand things, and these chunks of data may be greater or smaller depending on how much data we can assimilate at one time. Here we associate more information with the greater number of and greater size chunks of data.

Data sets are either fine-grained or coarse-grained, a qualitative distinction determined by the various descriptions of the process we employ. Just like sandpaper, coarse-grained means that there are only a few large grains of sand per square centimeter of the paper, and this paper can be used to remove large, obvious features from wood. On the other hand, fine-grained means that there are a great many very small grains of sand per square centimeter of paper, and the paper's surface may appear smooth to the touch. This sandpaper is used to polish the finest detail of a carved surface. The analogy carries over to the refinement of our understanding regarding a complex network. Fine-grained categories might include classification in school by the number of six-week periods completed, age in months or days and salaries in cents rather than dollars. More coarsely grained measures might include grade category such as elementary, middle or high school, age classification as early childhood, teenage and adult, along with salaries being in the four or five zero range. What constitutes a coarse-grained or fine-grained description of a process or phenomenon depends on what the natural scales are for the process, versus what is being investigated and what it is we want to learn.

In some cases the decision about the level of detail in data is arbitrary and of no substantive significance, whether the height of a tsunami is measured in meters or feet is probably not very important. In other cases the decision to work with either the fine- or coarse-grained data can be quite important in developing information about the process under investigation. As a rule of thumb, always measure at the most fine-grained level that is feasible, and then subsequently, if necessary or desirable, aggregate the data to obtain a more coarse-grained description.

2.2.2   Representing the data

Typically time series are displayed in the form of a graph, with the magnitude of the dependent variable on the vertical axis and time, the independent variable, on the horizontal axis. The data points are the values of the measurements of the phenomenon of interest, whether they are test scores for disadvantaged students in the inner city for each of the past ten years, the number of deaths due to AIDS in the decade of the eighties or the number of births to teens since the sixties, these are all time series. Each time series tells a story, but in a language that we must relearn with each new phenomena and with each retelling of the story.

There is a vast array of measures that one can apply to various time series to extract the structure of interest, but often the simple application of these measures is not enough. Rather one must have a strategy that is tailored to the phenomena in which one is interested, that is, one must develop a way of transforming the fluctuating data into reliable patterns. The measures are the tools; the time series is the marble or granite, and whether one winds up with a *David* or a pile of white chalk depends on a combination of talent, training, luck and the willingness to work hard.

How do we decide what data regarding an individual's physiology should be recorded and displayed? Centuries of clinical medicine has determined that during a complete physical examination the physician needs to know an individual's heart rate, blood pressure, whether or not s/he has trouble walking and so on. These are the traditional indicators of health dating back to the nineteenth century and the introduction of the notion of homeostasis into medicine. It is evident, however, that these indicators are all average quantities. But is it true that the best diagnostic indicator of the status of the cardiovascular network is the heart rate? Can't we do better than the breathing rate to determine the working state of the respiratory network? Is the variability in these and the other complex physiological networks we monitor actually best characterized by the bell-shaped distribution of Gauss? The answer to these and other similar question lies in the data [187].

Figure 2.7a displays what has come to be called heart rate variability (HRV). The data consists of the time interval from one beat to the next in 100 consecutive heart beats taken from a much longer sequence, where a young healthy adult is lying supine. It is evident from the figure that the heart rate is not constant. The average time interval between beats is approximately 0.6 sec, which would yield a heart rate of 100 beats per minute, if the beat-to-beat time interval did not change over time. However, although the time interval does change, it is not a radical change. The variation in heart rate is approximately 10%, as measured by the standard deviation of 0.06 sec, and this value seems to support the hypothesis of normal sinus rhythm, since 0.06 sec is very much less than 0.60 sec. After all, how much structure could be contained in these apparently random changes in the time intervals between beats?

The time series in Figure 2.7a looks like a sequence of random numbers, hopping between values without rhyme or reason. But we can do a simple operation on the data, that is, randomly shuffle the time intervals, to see if there is any underlying pattern. Shuffling destroys any influence between successive heart beats that exist in the original data set. This test for structure is depicted in Figure 2.7b, where the same average value and standard deviation is obtained as in the original data set because no time intervals have been added or subtracted, the data points have only been randomly moved around relative to one another.

Visually comparing Figure 2.7a with 2.7b we can see that certain patterns of time intervals monotonically increasing or decreasing over short time intervals are suppressed. No such regularity in the time interval data survives the shuffling process in Figure 2.7b. We can see from the data shown, which is typical of normal healthy people, that characterizing the cardiac cycle with a single heart rate ignores any underlying patterns of change within the data. Patterns contain information about the process being examined, so by ignoring such patterns in physiologic time series, information is discarded that may be of value to the physician. In fact it is quite possible that the

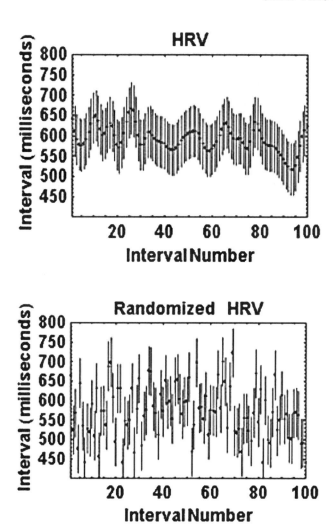

Figure 2.7: [a] (Upper) The time series for the interbeat intervals of a normal healthy individual lying supine are given by the dots and the bars indicate plus and minus the standard deviation. [b] (Bottom) The same data as above, but randomized with respect to the beat number. The comparison of the two time series shows that characterizing the beating heart by a single number, such as an average heart rate, appears to be inadequate [from [187] with permission].

information contained in the variability pattern may be more important than the average values used in setting the parameters in an implanted cardiac pacemaker.

A pattern in the HRV time series can be extracted in a simple manner by determining if structure is contained within the time series. If the HRV data has structure, often called correlations in a statistics context, this implies that what happens at a given point in time influences what happens at a later time. A graphical way to reveal this dependence is by determining how the average and standard deviation of the HRV data change in time and then graphing the two measures against one another. This is shown in Figure 2.8. All the structure is quantified by the scaling exponent, which is the slope of the standard deviation versus average curve. An uncorrelated random process would have a slope of $H = 0.50$, a slope greater than this indicates a long-time memory in the time series. The correlation contained in the HRV time series in Figure 2.7 is revealed in Figure 2.8. The shuffled data from Figure 2.7b is seen to have a slope very nearly equal to one-half, from which it is safe to say any structure in the original HRV data is due to the time ordering of the data points. The slope of the original HRV time series in Figure 2.7a is $H = 0.96$ and being greater than $H = 0.5$ implies a positive correlation. This positive correlation in the HRV data means that in the dynamics of the heart a long time interval is probably followed by an even longer time interval and a short time interval is probably followed by an even shorter time interval. This statistical organization of time intervals is the phenomena of persistence in the heart rate variability. The persistence is characteristic of a normal healthy heart.

The HRV index obtained for the HRV time series is a measure of the complexity of the cardiovascular network and has a well-defined range for normal healthy individuals. The HRV index changes as we age, becoming closer to one-half the older we get, indicating the loss of coherence in the generation of time intervals. On the other hand the HRV index approaches one as the variability in heartbeat is lost.

This example is only the tip of the iceberg. The argument used for the HRV time series has also been applied to the respiratory network using breathing rate variability (BRV) time series for the time intervals between breaths; the motorcontrol network using stride rate variability (SRV) time series for interstride intervals; the regulatory network for body temperature using body temperature variability (BTV) time series; and finally gastrointestinal rate variability (GRV) using time intervals between the contractions

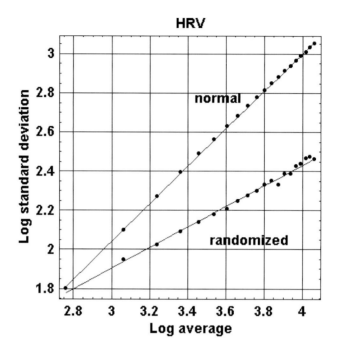

Figure 2.8: The information generated as a function of time by the heart beat interval time series or HRV for a typical time series has a slope of $H = 0.96$, consistent with the scaling index obtained by other scientists using a variety of techniques. This information curve is compared with that generated by the same data shuffled to suppress the long-time correlation, yields a slope of $H = 0.54$. The results clearly demonstrate that a traditional measure such as the average heart rate is an inadequate measure of health because the two time series shown in Figure 2.7 have the same average value [from [187] with permission].

of the stomach during digestion. Each of these complex networks presents a scaling index as a better indicator of the state of health of the underlying network than does the associated average value, see West [187] for a complete discussion and references to the appropriate literature.

2.3 Fractal statistics

We have avoided using the term signal in connection with time series because signal implies data containing information about the complex phenomena being interrogated. Signals come in many shapes and sizes depending on the network. Consequently, to make the discussion manageable we restrict the data sets discussed here to those dealing with the function and activities of life and of living matter such as organs, tissues and cells, that is, to physiology. Complexity in this context incorporates the recent advances in physiology concerned with the application of concepts from fractal statistics to the formation of a new kind of understanding in the life sciences.

However, even physiological time series are quite varied in their structure. In auditory or visual neurons, for example, the measured quantity is a time series consisting of a sequence of brief electrical action potentials with data regarding the underlying dynamical system residing in the spacing between spikes, not in the pulse amplitudes. A very different kind of signal is contained in an electrocardiogram (ECG), where the analogue trace of the ECG pulse measured with electrodes attached to the chest, reproduces the stages of the heart pumping blood. The amplitude and shape of the electrocardiograph (ECG) analogue recording carries information in addition to the spacing between heartbeats. Another kind of physiological time series is an electroencephalogram (EEG), where the output of the channels attached at various points along the scalp, recording the brain's electrical potential, appears to be random. The information on the operation of the brain is buried deep within the data in the form of erratic fluctuations measured at each of these positions along the scalp. Thus, a biological signal, here a physiological time series, can have both a regular part and a fluctuating part; the challenge is how to best analyze the time series from a given physiologic network to extract the maximum amount of information.

Biological time series have historically been fit to the signal plus noise paradigm used in engineering. The signal is assumed to be the smooth, continuous, predictable, large-scale motion in a time series. Signal and pre-

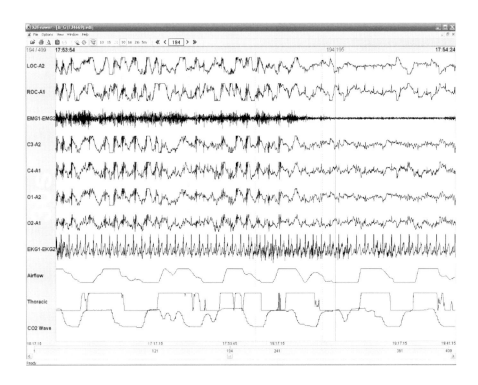

Figure 2.9: Typical example of physiological time series for a patient. The figure shows several EEG signals from different locations, ECG signal (EKG1-EKG2), airflow and CO_2 wave from respiration, and thoracic movements.

dictability go together, in that signals imply information and ultimately knowledge, and very often the mechanistic interpretation of information has to do with our ability to associate that information with mechanical, predictable processes within the network generating the time series. Noise, on the other hand, is typically discontinuous, small-scale, erratic motion that is seen to disrupt the signal. The noise is assumed, by its nature, to contain no information about the network, but rather to be a manifestation of the influence of the unknown and uncontrollable environment on the network's dynamics. Noise is considered to be undesirable and is removed from the time series by filtering whenever possible. The mathematician Norbert Wiener, as his contribution to the war effort during World War Two, gave the first systematic discussion of this partitioning of erratic time series into *signal* and *noise* in his book *Time Series* [190], using the then newly extended theory of harmonic analysis applied to communication theory. Wiener's partitioning of effects does not take into account the possibility that the underlying process can be complex and such complexity does not allow for this neat separation into the traditional signal and noise components. Complexity is, in fact, the usual situation in social and life sciences so we ought not to expect that time series separate in general. The signal plus noise paradigm does not apply to time series from complex networks; complexity that is often manifest in the fractal properties of the time series.

The fractal concept was formally introduced into science by Beniot Mandelbrot over twenty years ago and his monograph [102] brought together mathematical, experimental and disciplinary arguments that undermined the traditional picture of the physical world. Since the early eighteenth century it had been accepted that celestial mechanics and physical phenomena in general are described by smooth, continuous, and unique functions. Mandelbrot called into question the accuracy of the traditional perspective of the natural sciences, by pointing to the failure of the equations of physics to explain such familiar phenomena as turbulence and phase transitions. In his books [102, 103] he catalogued and described dozens of physical, social, and biological phenomena that cannot be properly described using the familiar tenants of dynamics from physics. The mathematical functions required to explain these complex phenomena have properties that for a hundred years had been categorized as mathematically pathological. Mandelbrot argued that, rather than being pathological, these functions capture essential properties of reality and are therefore better descriptors of the physical world's data than are the traditional analytical functions of nineteenth century physics and engineering.

Living organisms are immeasurable more complicated than inanimate objects, which partly explains why we do not have available fundamental laws and principles governing biological phenomena equivalent to those in physics. For example, there are no biological equivalents of Newton's Laws, Maxwell's equations and Boltzmann's Principle for physiological phenomena. Schrödinger [151] laid out his understanding of the connection between the world of the microscopic and macroscopic, based on the principles of equilibrium statistical physics. In that discussion he asked why atoms are so small relative to the dimensions of the human body. The answer to this question is both immediate and profound. The high level of organization necessary to sustain life is only possible in macroscopic networks; otherwise the order would be destroyed by microscopic (thermal) fluctuations. A living organism must be sufficiently large to maintain its integrity in the presence of thermal fluctuations that disrupt its constitutive elements. Thus, macroscopic phenomena are characterized by time averages; averages that smooth out microscopic fluctuations. Consequently, any strategy for understanding physiology must be based on a probabilistic description of complex phenomena, and as we shall see, on an understanding of phenomena lacking characteristic time scales, that is, on self-similar or fractal scaling.

There are different types of fractals that appear in the life sciences; see Figure 2.10. For example we have geometrical fractals, that determine the spatial properties of the tree-like structures of the mammalian lung, arterial and venous systems, and other ramified structures [182]; statistical fractals, that determine the properties of the distribution of intervals in the beating of the mammalian heart, breathing, and walking [183]; finally, there are dynamical fractals, that determine the dynamical properties of networks having a large number of characteristic time scales, with no one scale dominating [184]. In the complex networks found in physiology the distinction between these three kinds of fractals blur, but we focus our attention on the statistical rather than the geometrical fractals, in part, because the latter have been reviewed in a number of places and do not directly relate to our present concerns [108].

The historical view of complexity involved having a large number of variables, each variable making its individual contribution to the operation of the network and each variable responding in direct proportion to the changes in the other network variables. Small differences in the input could be washed out in the fluctuations of the observed output. The linear additive statistics of measurement error or noise is not applicable to the complex net-

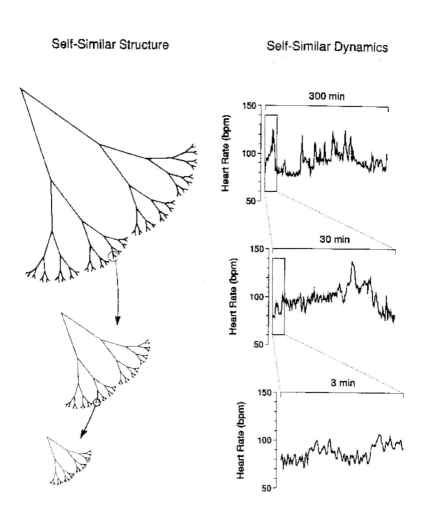

Figure 2.10: Left, schematic of a tree-like fractal has self-similar branchings such that the small scale (magnified) structure resembles the large scale form. Right, a fractal process such as heart rate regulation generates fluctuations on different time scales (temporal "magnifications") that are statistically self-similar. (From http://www.physionet.org/; adapted from [59]).

works of interest here. The elements in complex physiologic networks are too tightly coupled, so instead of a linear additive process, nonlinear multiplicative statistics more accurately represent the fluctuation, where what happens at the smallest scale can and often is coupled to what happens at the largest scale. This coupling is manifest here through the scaling index.

To paraphrase West and Griffin [186], the signal plus noise paradigm used by engineers is replaced in physiology with the paradigm of the high wire walker. In the circus, high above the crowd and without a net, the tightrope walker carries out smooth slow motions plus rapid erratic changes of position, just as in the signal plus noise paradigm. However, the rapid changes in position are part of the walker's dynamical balance; so that far from being noise, these apparently erratic changes in position serve the same purpose as the slow graceful movements, to maintain the walker's balance. Thus, both aspects of the time series for the wirewalker's position constitute the signal and contain information about the dynamics. Consequently, if we are to understand how the wirewalker retains balance on the wire, we must analyze the data, that is, the wirewalker's fine tuning to losses of balance as well as the slow movements. This picture of the wirewalker more accurately captures the view of signals generated by complex networks than does the signal plus noise paradigm.

The individual mechanisms giving rise to the observed statistical properties in complex networks are very different, so we do not even attempt to present a common source to explain the observed scaling in walking, breathing and heart beating. On the other hand, the physiological time series in each of these phenomena scales in the same way, so that at a certain level of abstraction the separate mechanisms cease to be important and only the relations matter and not those things being related. It is the relation between blood flow and heart function, between locomotion and postural balance, between breathing and respiration, which are important. If a science of networks is to be realized then it must be the case that such relations often have a common form for complex networks. This assumption is not so dramatic as it might first appear. Consider that traditionally such relations have been assumed to be linear, in which case their control was assumed to be in direct proportion to the disturbance. Linear control theory has been the backbone of homeostasis, but it is not sufficient to describe the full range of HRV, SRV and BRV. Traditional linear control theory cannot explain how the statistics of time series become fractal, or how the fractal dimension changes over time.

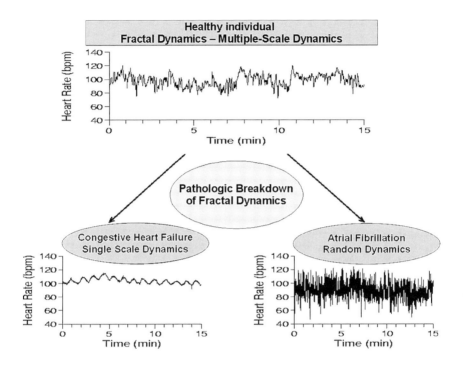

Figure 2.11: Breakdown of a fractal physiological control mechanism. The upper HRV time series is for a normal healthy individual with a fractal dimension between 1.1 and 1.3. The lower left HRV time series is that of an individual undergoing congestive heart failure with a fractal dimension near 1.0. The lower right HRV time series is for a person having an atrial fibrillation with a fractal dimesion near 1.5. (From http://www.physionet.org/; Adapted from [59]).

The fractal dimension of a process is related to the scaling index, such as the one depicted in Figure 2.11. This dimension can be very useful in determining how well a network functions, for example, the cardiovascular network. In Figure 2.11 is depicted the time series for the HRV of three different individuals. It is clear from the figure that the three time series have the same average heart rate. The individual at the top of the figure is healthy with a HRV fractal dimension (scaling index) in the normal range. The person on the lower left has a fractal dimension of nearly one and is dying of congestive heart failure. It is clear that the time series for this latter individual is more ordered than that for a healthy person and s/he is dying by being overly organized. The HRV of the third sequence on the lower right has a fractal dimension (scaling index) close to three-halfs. The time series is more random than that of a healthy person and this individual is dying by being overly disorganized because of atrial fibrillation. In short, the average heart rate does not distinguish among the three cases, but the fractal dimension (scaling index) does. Thus, in the real world the complexity of a healthy physiologic network is captured by the scaling or variability index, not by an average.

2.4   Solar and climate variability

The simple fact that the sun warms the earth suggests a direct connection between the Earth's globally averaged temperature and solar activity influencing the total solar irradiance. Consequently, any significant changes in solar activity should result in corresponding changes in the Earth's global temperature. The literature concerning the solar influence on the Earth's temperature is quite extensive. On the other hand, volcanic activity and human pollutants also effect climate. Thus, according to the traditional scientific approach, the accepted strategy for understanding global temperature has been to dissect the phenomenon of global warming, separate the various hydro-thermal mechanisms and construct large-scale computer simulation models, such as the general circulation model implemented at the NASA Goddard Institute for Space Studies [62]. The goal is to estimate the relative sizes of the individual contributions to global warming. The findings have been summarized in the IPCC report, as we discussed in Chapter 1.

However, it is well established that both the Sun and the Earth are complex networks as it is manifest in the complex and apparently random fluctuations in solar activity (for example, in the number of solar flares and sun spots) and the Earth's global temperature anomalies. The dynamics of the Sun's surface is turbulent, as is evidenced by changes in solar flare and sun spot activity, as well as strong erratic fluctuations associated with the intermittency of those activities [50, 61, 16, 57], and the changes in TSI. Then, the time variation in the solar activity induces changes in the Earth's average temperature through multiple mechanisms which can be sensitive to variations of total solar irradiance, ultra-violet (UV) radiation and modulation of cosmic rays [50]. The Sun's influence on climate change produces trends that move the Earth's global temperature up or down for tens or even hundreds of years.

The Earth's atmosphere and oceans are also turbulent complex networks: variations in solar activity represent the input to these networks while the average global temperature is an output. However, when a complex network has to analyze an input the response may not be simply related to the input signal, unlike a simple linear relation that can be evaluated with an opportune cross-correlation analysis. In general the input and the output data may appear to have similar strongly correlated patterns (and in this case the linking is evident to all) or they may appear very different and uncorrelated (and in this case most scientists would conclude that a linking is unlikely). Linear regression methods have been developed to evaluate the intensity of the linking in these situations. But when a complex network processes an input signal, the response may not be so simply predicted. A strange situation sometime occurs: a strong linking may indeed exist, but it may be complex and nonlinear, and for this reason simple multi-linear regression methods fail to identify it.

The Sun-Climate linking has evidently all the credentials of manifesting itself according to the properties of the above latter case: the complexity case. Indeed, numerous studies have claimed a correlation between solar variations and climate variability. This has happened since antiquity, for example in the fourth century B.C. Theophrastus reported that when the sun has spots, the weather tends to be wetter and rainier [66]. Since the nineteenth century when a rigorous sunspot number record was finally available,[4] scientists discovered several evident correlations between solar dynamics and the Earth's climate. But problems arose as well. As Hoyt and Schatten say in their book

---

[4]The sunspot number has been systematically recorded since Galileo Galilei started observing them with his telescope in 1610.

The Role of the Sun in Climate Change [66]:

> ... dozen of papers appeared relating changes in the sun to variations in the Earth's temperature, rainfall and droughts, river flow, cyclones, insect populations, shipwrecks, economic activity, wheat prices, wine vintages and many other topics. Although many independent studies reached similar conclusions, some produced diametrically opposed results. Questions critics asked included: Why were people getting different answers at different locations? Why did some relationships exist for an interval and then disappear? Were all these results mere coincidences?...

Scientists were and are still divided about how to interpret the Sun-climate linking results. Does the problem reside in the data? Or does the problem reside in the interpretation of the data? Or, finally, does the problem reside in the fact that our knowledge of this phenomenon is based on computer simulations that may mislead us?

We discuss the above points subsequently, but right now we believe that it is a well established fact that both the Sun and the Earth are complex networks as measured by random fluctuations in solar activity and global temperature anomalies. As we explained early in this chapter, a manifestation of complexity occurs when the statistics of the fluctuations in the physical observables are described by an inverse power law rather than by a Gaussian distribution. We [140, 143, 144] have determined that both solar and climate records manifest *complexity* according to the above inverse power law statistical requirement. Moreover, it has recently been determined that when one complex network perturbs another complex network that the latter adopts the statistics of the former under certain circumstances [5], this is an information transfer process. Indeed, a few years ago, we demonstrated that the conditions found to be necessary for this kind of transfer to occur are met in the Sun-climate network [140, 143, 144].

This statistical affinity between the fluctuations in TSI and the Earth's average temperature suggests the existence of a nonlinear complex linking between the two. In fact, the probability that two networks manifest the same anomalous statistics as a mere coincidence is quite slim. The nonlinear linking may also explain why linear methods have failed to give a definitive answer regarding the relative importance of the various causes of global warming. But, right now we introduce the solar and climate data, and uncover additional problems that make the entire debate difficult even from a traditional point of view.

2.4.1 Solar data

In this section we introduce the solar data used in the computer calculation of global warming. Two of the questions to be answered regarding the data are: Which data do we use? How reliable are the data we use?

Since ancient times Mesopotamian, Egyptian, Greek, Chinese and medieval astronomers have occasionally observed dark spots appearing and disappearing on the surface of the sun. Sunspots appear as dark spots on the Sun's surface because these areas are cooler than the surrounding regions. However, the first collection of solar data started in the first half of the seventeenth century when Galileo Galilei (1564-1642) and other scientists could observe the Sun's surface with the help of the newly invented telescope. Galileo observed that these spots not only could appear and disappear but that they were moving on the Sun's surface in an almost regular way and concluded that the sun revolved on its own axis: a fact that was not previously known and was in contrast with the ancient Aristotelian cosmology that claimed the immutability of the celestial bodies. Thus, Galileo's discovery proved the existence of a solar dynamics and several scientists after him were interested in systematically collecting solar data for determining how the Sun's activity evolves in time.

In 1843 an amateur astronomer, Samuel Heinrich Schwabe (1789-1875), after 17 years of careful observations of the sun noticed that the average number of sunspots changes periodically from year to year. The solar cycle was the first major discovery concerning solar dynamics and motivated Rudolf Wolf (1816-1893) to compile a reconstruction of the sunspot number back to 1745 and, eventually, back to the earliest sunspot observations by Galileo and contemporaries. Because sunspots come in many shapes, sizes and different levels of grouping, the early solar astronomers have found it useful to define a standard sunspot number index, which continues to be used today.

The average duration of the sunspot cycle is 11 years, but cycles are not perfectly periodic nor equal in amplitude. The sunspot cycle period has been observed to vary from 8 to 14 years, and large variations in amplitude are observed as well. Edward Walter Maunder (1851-1928) also noticed that the sunspot cycle can disappear for a long time period as occurred from 1645 to 1715. This period is now known as the Maunder Minimum and it is quite interesting because during that period European and Asian peoples recorded extraordinarily cool winters. Thus, the solar Maunder Minimum seems to be associated with a climate period know as the Little Ice Age.

In 1852, within a few years after Schwabe's discovery, Edward Sabine (1788-1883), Rudolf Wolf, Jean-Alfred Gautier (1793-1881) and Johann von Lamont (1805-1879) independently and more or less simultaneously announced that the sunspot cycle period was "absolutely identical" to that of the Sun's geomagnetic activity, for which reliable data had been accumulated since the mid-1830s and partial data covering a few years were collected in Paris and in London since 1780: see Figure 2.12. This discovery marked the beginning of solar-terrestrial interaction studies.

Thus, all the above observations suggest that solar activity varies in time, but can this variation be truly linked to climate change? Evidently, a change in sunspot number on the solar surface may be of very little importance for the Earth's climate if it is not linked to other solar phenomena such as, for example, the total solar irradiance that can be easily proven to physically influence climate. However, before Schwabe's discovery, scientists believed that the solar luminosity was constant and several attempts to measure it were made. In 1838 the concept of a *solar constant*, as referring to the solar luminosity, was introduced by the French physicist Claude Pouillet (1790-1868) and the British astronomer John Herschel (1792-1871). The original measured value of the solar constant was about half the accepted modern value of 1367 ± 4 W/m^2 because these scientists failed to account for the absorption by the Earth's atmosphere. Better measurements were repeated later at different altitudes, such as the ones carried out by the American scientist Samuel Langley (1834-1906). However, the belief that the solar luminosity was *constant* was commonly held until a few decades ago and obstructed the development of a solar-climate interaction science during the twentieth century.

The concept of a *solar constant* was abandoned only when direct detailed measurements of the solar luminosity became possible with the advent of satellites. Observations from satellites established that solar luminosity undergoes a cycle very similar to that observed in the sunspot number record. Since then, attempts to reconstruct solar luminosity dynamics have been made. However, due to the intrinsic complexity of the network of physical mechanisms within the Sun, some technical difficulties as well as a significant dose of bad-luck, have lead to several problems as well as some controversies during the last decade. Let us see what happened.

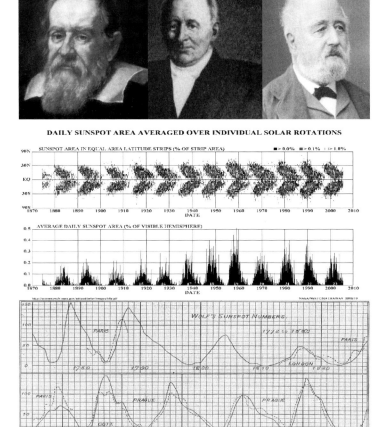

Figure 2.12: Top: Galileo Galilei (1564-1642), Samuel Heinrich Schwabe (1789-1875) and Rudolf Wolf (1816-1893). Middle: Sunspot number and area. Bottom: a nineteenth century comparison between the Wolf sunspot number and the geomagnetic activity index (Diagram reproduced from A.C. Young's *The Sun* revised edition, 1897).

TSI observations have been made by contiguous and overlapping satellite experiments since 1978 during solar activity cycles 21-23 (1980-2002): see Figure 2.13. The solar activity minimum marking the inception of solar cycle 24 has been expected in 2007/9. Two types of experiments have provided TSI data: self-calibrating, precision TSI monitors and Earth radiation budget (ERB) experiments.[5] The optimum composite TSI time series utilizes TSI monitor results to the maximum extent possible. However a two year gap in the TSI monitoring record between the ACRIM1 and ACRIM2 experiments (1989-1991) would have prevented compilation of a continuous record over the period from 1978 to date were it not for the availability of ERB results during the gap. This gap exists because NASA delayed the ACRIM2 mission following the 1986 Challenger space shuttle disaster. The relationship between ACRIM1 and ACRIM2 results across the ACRIM gap can be derived only using the overlapping ERB data sets which are less precise than the other records.

During the ACRIM gap the two ERB data sets behave in different ways: Nimbus7/ERB shows a slight upward trend, while ERBS/ERBE shows a slight downward trend. So, at least two alternative TSI satellite composites are possible and both have been suggested in the literature. The ACRIM TSI composite [193] uses the Nimbus7/ERB ACRIM gap ratio and exhibits a 0.04 percent per decade trend between the minima of cycles 21 to 23. A trend of this magnitude, sustained over many decades or centuries, could be a significant climate forcing. The ACRIM composite trend to higher TSI during cycles 21-23 and then back to approximately the level of the cycle 21-22 minimum may turn out to be the signature of a 22-year Hale cycle. Alternatively, the PMOD composite [52] uses the ERBS/ERBE satellite TSI database to calculate the calibration between ACRIM1 and ACRIM2, and modified both Nimbus7/ERB and ACRIM1 results based on the predictions of TSI proxy models and finds no significant trend over the cycle 21 to cycle 23 period.[6]

---

[5]Self-calibrating, precision TSI monitors include SMM/ACRIM1, UARS/ACRIM2, SOHO/VIRGO, ACRIMSAT/ACRIM3 and SORCE/TIM experiments and Earth radiation budget include Nimbus7/ERB and ERBS/ ERBE experiments. TSI monitors provide much higher accuracy and precision and are capable of self-calibrating the degradation of their sensors. The ERB experiments are designed to provide less accurate and precise TSI boundary value results for ERB modeling and cannot self-calibrate sensor degradation.

[6]It may be argued that TSI proxy models are not competitive in precision or accuracy even with the lowest quality satellite TSI observations. Their use in constructing the

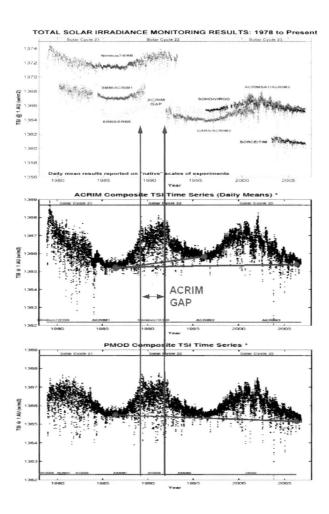

Figure 2.13: Top: The seven total solar irradiance monitoring results since 1978. Middle: the Active Cavity Radiometer Irradiance Monitor (ACRIM) TSI composite. Bottom: the Physikalisch-Meteorologisches Observatorium Davos (PMOD) TSI composite of the TSI satellite data shown on the top.

Thus, ACRIM more faithfully reproduces the best published satellite data, that is, the observations. On the other side, PMOD shows a pattern that is closer to some TSI proxy such as the record of the number of sunspots. We do not intend to critique these two results here, that is something that is more appropriately done in a technical journal and is part of the ongoing discussion in the scientific community [149]. Thus, ever since the total solar irradiance could be directly measured there is disagreement on whether the solar activity experienced an increase from solar cycle 21-22 (1980-1991) to solar cycle 22-23 (1991-2002) followed by a decrease during solar cycle 23-24 (2002-2013) as ACRIM would suggest or, alternatively it experienced a slight decrease since 1980 as PMOD would suggest: see Figure 2.13. It is evident that the two composites have different implications about whether and how much the Sun contributed to the global warming observed since 1980.

If there are controversies about the solar luminosity evolution during the last three decades, when direct measurement of TSI has been possible, a reader can expect that even more significant uncertainties are to be found about the solar activity reconstructions of the data from past centuries and millennia when no direct measurement of TSI was possible. Indeed, several proxy reconstructions of solar activity have been suggested during the last decade: see Figure 2.14. These reconstructions are based on several proxies supposed to manifest the total solar activity to some degree. One proxy of special importance is the sunspot record (sunspot number [158] and sunspot cover [133]). Their temporal variation can be associated with equivalent variation of the solar magnetic activity. The sunspot record clearly shows the 11-year quasi-periodic Schwabe solar cycles and that the amplitude as well as the average trends of these cycles are not constant in time. Two consecutive 11-year Schwabe solar cycles give one 22-year Hale solar magnetic activity cycle. Longer solar modulation activity reconstructions can be inferred from

---

PMOD composite convolutes the relatively high uncertainties of the models with the more precise observational data and conforms the PMOD composite to the proxy model rather than providing the most valid interpretation of the TSI observational database. Moreover, the absence of a trend in the PMOD composite and any composite based on the ERBS/ERBE ACRIM gap ratio seems to be an artifact of uncorrected degradation of ERBS/ERBE results during the gap. The modification of Nimbus7/ERB and ACRIM1 results are particularly evident during the solar maxima periods 1978-1980 and 1989-1991. However the experimental groups, who have published the satellite data, disagree about the necessity of the modifications implemented by the PMOD group and argue that it is more likely that the proxy reconstructions fail to correctly reproduce the solar maxima.

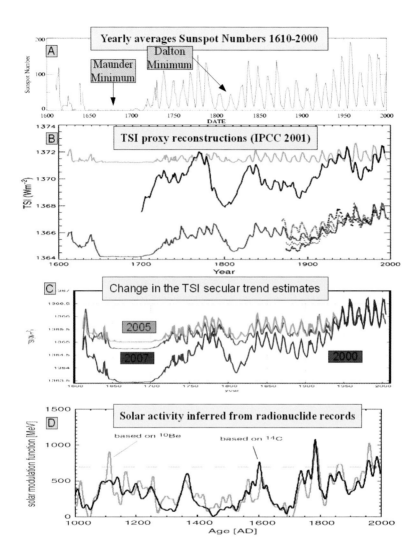

Figure 2.14: [A] Sunspot number record. [B] Several reconstructions of TSI available in 2001, (red [83], black [67], and others for example [91, 160]). [C] Three recent TSI proxy reconstructions proposed in different years (blue [84], green [178], red [77]). [D] Two recent solar modulation activity reconstructions since 1000 AD inferred from ^{10}Be and ^{14}C radionuclide records [117].

^{10}Be and ^{14}C radionuclide records that are supposed to be an indirect measure of the solar magnetic activity because their production is related to the cosmic ray flux that is modulated by the strength of the solar magnetic field.

The secular variation of the sunspot cycles and of the radionuclide records imply that the total solar activity is clearly not constant, but evolves in time. Even the most ardent opponent of the argument for the solar contribution to global warming would be reluctant to dismiss the fact that there exists a solar dynamics that may influence climate. However, as Figure 2.14 shows there is no agreement on how large the variation of the solar activity is. Note that the most recent TSI proxy reconstructions, such as those shown in Figure 2.14C [178, 77] show a secular variability significantly smaller than that shown by previous reconstructions which are shown in Figure 2.14B.

A quantitative measure is necessary to estimate the solar contribution to global climate change by means of traditional computational climate models. So, because the quantitative aspect is uncertain the model predictions are necessarily uncertain as well. It is evident that a model predicts a larger or smaller solar effect on climate change according to the magnitude of the estimated secular variation of the TSI over the centuries. It is also evident that because the latest TSI proxy reconstructions present a smaller variability than the previous one, if a model produced a good fit with the temperature record by using the former reconstruction, the same model would not fit the same temperature data if forced with the latter TSI proxy reconstructions. Thus, all climate model studies adopting the old TSI proxy reconstructions and claiming good correspondence with global temperature data should be considered suspect because the estimated solar forcing has changed since these calculations were done. Note that the general circulation model implemented by the NASA Goddard Institute for Space Studies [62], shown in Figure 1.16, adopts as solar forcing that deduced from the blue TSI reconstruction shown in Figure 2.14C and constructed in 2000. By using a more recent TSI reconstruction, such as the green that was constructed in 2005, the model would predict approximately one third of the warming observed during the period 1900-1950. Thus, this prediction would result in the computer model no longer fitting the data and suggests that the computer model needs to be revised.

Nevertheless, Figure 2.14 also shows that from a qualitative point of view all reconstructions look similar. In fact, they present equivalent 11-year solar cycles modulated by equivalent cooling and warming periods, such as during the Maunder minimum (1645-1715) and Dalton minimum

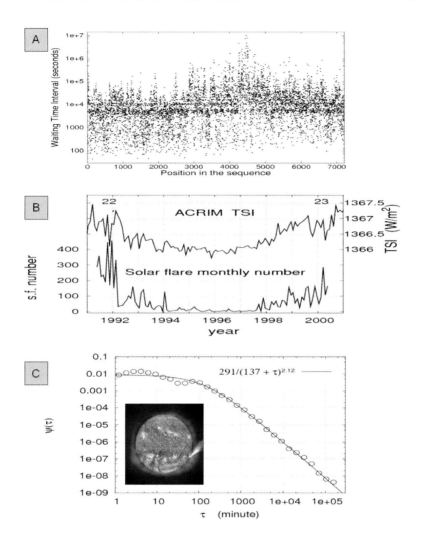

Figure 2.15: [A] Waiting-time intervals between consecutive Hard x-ray solar flares between 1991 to 2001. [B] Comparison between monthly number of solar flares and ACRIM composite TSI monthly average during the solar cycle 22-23. [C] The waiting time distribution between Hard x-ray solar flares is an inverse power law with Pareto exponent $\mu = 2.12$ (from [61, 143] with permission).

(1780-1840), and show a steady increase of activity since the middle of the nineteenth century. Thus, it is legitimate to ask whether this qualitative pattern of the solar dynamics, which is more certain than its quantitative properties, may be used to better evaluate the Sun's influence on climate change. We have found that the latter leads to an empirical approach to climate change, which is an alternative to the computational climate model approach.

If solar dynamics has to be evaluated from a statistical point of view, an important solar record to take into consideration is that of solar flares. Solar flares are extremely violent explosions on the Sun's surface during which enormous amounts of energy are emitted as UV and X rays, and as flux of atomic particles. The UV, X-ray, solar wind modulations and magnetic storms induced by solar flare intermittency are known to cause strong perturbations in the Earth's ionosphere. These perturbations alter radio-communications and cause aurora [50]; they may effect global climate as well by altering the ion and photon chemistry of the atmosphere [66, 140, 143]. Solar flares are important because their intermittent nature enables us to identify an alternative solar measure directly linked to the turbulent dynamics of the Sun [61, 143].

Figure 2.15 shows some properties of the Hard X-ray solar flare occurrence record covering approximately 10 years from 1991 to 2001. Figure 2.15B shows that the solar flare rate is closely related to the TSI record. The two records apparently track one another rather closely and there is even alignment between several spikes in both records. This behavior suggests that the solar flare data may, in some cases, be used as a surrogate for the TSI time series. Figure 2.15C shows that the waiting-time distribution[7] between hard X-ray solar flares is an inverse power law with Pareto exponent $\mu = 2.12$: a fact indicating that solar flare dynamics can be described by Lévy statistics, and not by a Gaussian distribution. The above two findings suggest that the entire solar dynamics can be generated by a process that is statistically affine to a Lévy process. Thus, although the exact amplitudes of the trends of the solar dynamics is uncertain, the fact that such a dynamics obeys an anomalous statistics raises the question as to whether the Earth's climate adopts the same statistics by means of an information transfer process.

---

[7]A waiting-time distribution is the relative frequency of time intervals of a given size between event, in this case the event is a solar flare.

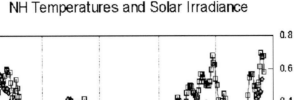

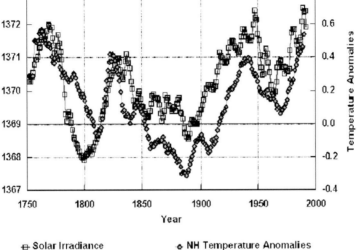

Figure 2.16: A simple superimposition between Hoyt and Schatten's Total Solar Irradiance reconstruction and a Northern Hemisphere average temperature reconstruction [67].

2.4.2 Temperature data

There are several methods that can be used to detect a solar signature in global temperature data and we discuss some of them in later chapters. We point out that it is always possible to argue that finding a common pattern between two time series may be accidental and, therefore, such a finding may not imply a causal relation between the two phenomena. Indeed, when solar and climate secular reconstructions are compared an evident correlation appears during both cooling and warming periods since 1750, as Figure 2.16 shows, and figures like this one would suggest a strong solar effect on climate lasting since 1750. Climate model simulations too, apparently, fit the data well as Figure 1.16 shows. These large-scale computer models claim that the solar variation has only a minor effect on climate.

Evidently the above observation should be considered valid not only when the global temperature and solar data are compared, but also when global temperature and climate model simulation data are compared. Finding a simple correlation between two sequences may be misleading in both cases. Given the intrinsic complexity of climate, if a climate model produces a sequence that seems to fit the data, this may not imply that the model is correct, only that the model has been well tuned to fit the data during some time interval. Evidently, the only way to determine a supposed causality between two phenomena is when this causality is supported by several complementary analyses and persists for a long time period. However, there is an additional difficulty; global temperature data are significantly uncertain, and several different records referring to the same phenomena are available.

Figure 2.17 shows different observed surface and upper air global-mean temperature records (lower-mid-upper troposphere and lower stratosphere) since 1950. For each region the data sets are only slightly different due to statistical fluctuation. But if we move from lower to higher altitudes it is evident that the records shows different trends: the surface has warmed substantially, the stratosphere has cooled and the troposphere shows a gradual change in the trend from lower and upper regions. Also, it is evident that the troposphere is amplifying the temperature variations observed in the surface record.

Changes of trend depending on the altitude are expected and are in agreement with the actual understanding of the climate science. However, the data show particular patterns that are not understood yet nor reproduced by the computational models. From 1950 to 1980, the low and mid troposphere have warmed at a rate slightly faster than the rate of warming at the surface. This is in agreement also with the amplification effect observed in the troposphere for patterns at shorter time scales. However, since satellite measurements have been available (1979) the majority of data sets show that the warming at the Earth's surface is greater than in the troposphere: the troposphere presents only a slight warming over the last three decades.

It is not clear yet if this apparently anomalous behavior in temperature variation is due to the fact that some climate mechanisms are not well understood and, therefore, they are not implemented in the models yet. Alternatively, the satellite measurements of the tropospheric temperature are not well composed and the current records are underestimating the warming. A final option is that the surface temperature records are corrupted and the present records are overestimating the warming. One of the mechanisms

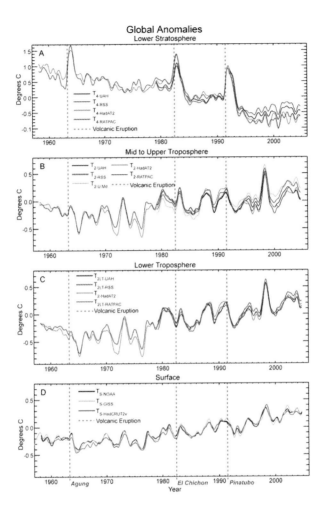

Figure 2.17: Observed surface and upper air global-mean temperature records (from [191] with permission). A) lower stratosphere records from two satellite analyses together with equivalently-weighted radiosonde records; B) mid- to upper-troposphere records from three satellite analyses together with equivalently-weighted radiosonde records; C) lower troposphere records from satellite and from radiosonde; D) surface instrumental records. All time series are based on monthly-mean data smoothed with a 7-month running average, expressed as departures from the Jan. 1979 to Dec. 1997 average.

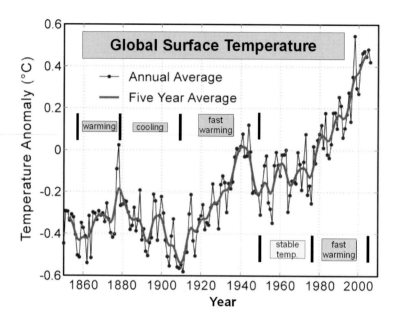

Figure 2.18: Global average instrumental temperature record since 1850 (re-drawn from [17]). Note the complex patterns made of consecutive cooling and warming periods. In particular, note the cooling from 1880 to 1910, the fast warming occurred from 1910 and 1950 and the almost stable value of the temperature from 1950 to 1975. Since 1975 the surface record shows significant warming.

that may explain a surface overestimation of the warming is the Urban Heat Island effect [107]. But we postpone discussion of this issue.

However, it is evident that it is necessary to have reliable data to test climate models and eventually forecast climate scenarios. Models should also be tested against long time series such as the secular temperature records. But here the data are much more uncertain.

Instrumental global surface temperature records, mainly based on direct thermometer readings, start from 1850 [17]. The percentage of the Earth's surface covered by this record increased in time. The last figure shows that the temperature underwent quite complex dynamics with alternate warming and cooling periods. There are two long periods of equally fast warming during the first and second half of the twentieth century. These two warming periods are interrupted by a 30 year period of stable temperature from 1945 to 1975. This non-monotonic patterns would be inconsistent with the monotonic increase of greenhouse gases in the atmosphere, which occurred during the same period associated with a progressive increase of industrialization.

Before 1850 there are not enough direct measurements of the surface temperature to reconstruct a global record. Past climate has been inferred from historical records and more recently by means of paleoclimate reconstructions that make use of several temperature proxies such as tree-rings, ice cores, corals, and sediments. There have also been attempts to reconstruct past climate from historical records. It is believed that during the medieval period from 800 to 1300 the Earth's climate experienced a significant warming while during the period from 1450 to 1850 the Earth's climate experienced a significant cooling: see Figure 2.19. After the mid-1800's, the climate of Europe has shown a dramatic warming trend that persists today.

The most compelling historical evidence for a Medieval Warm Period includes the fact that vikings colonized Greenland, and several agricultural factors. The same name *Greenland* indicates that land appeared to the Vikings "green" and likely there were some forests. Therefore, it is quite evident that during that period Greenland was warmer than today. Also, the fact that the Vikings could navigate the North Sea is evidence of a warmer climate because with cooler climates these seas are typically full of dangerous icebergs that would have made open-sea navigation almost impossible. Among the agricultural factors one is the extended cultivation of grape vineyards, which require moderate temperatures and a long growing season. During the Medieval Warm Period grape vineyards were found as far north as England while today grapes vineyards are typically only found as far north as France in Europe, approximately 350 miles south of the earlier location.

There are also historical evidences of a Little Ice Age during the period from 1450 to 1850. Vikings abandoned Greenland after the fourteenth century. The wine production in England and Germany gradually declined [80]. In the North of Europe, forests of beech trees were replaced first by oaks and subsequently by pines. General agricultural production significantly decreased effecting the economics and health of the European people. In particular, malnutrition led to a variety of severe and lasting illnesses aggravating influenza epidemics, agues and bubonic plagues; some fishery of the North of Europe collapsed because the fish stocks moved south and there was social unrest due to starvation. An extended study [118] of more than

12,000 paintings in 41 art museums in the United States and eight European countries revealed that during the epoch when the Little Ice Age is believed to be at its peak (1550-1850), cloudiness and darkness prevailed compared to the previous epoch (1400-1550) and the following epoch (1850-1967).

Also the climate effects of large volcanic eruptions are well established in human history. For example, there exists a "year without a summer" which followed the 1815 eruption of Tambora, in Indonesia. The eruption of the Laki volcano in Iceland in 1783 and the explosion of Krakatoa in the Sunda Strait between Java and Sumatra in Indonesia in 1883 also created volcanic winter-like conditions during summer. The massive Kuwae eruption between Epi and Tongoa islands in 1453 caused a sudden global climate cooling that may also have favored the fall of Constantinople. Byzantine historians recorded strange climate events such as on the night of May 22, 1453, one week before the collapse of the city, the moon, symbol of Constantinople, rose in dark eclipse, four days later, the whole city was blotted out by a strange thick fog unknown in May and the sky had a strange reddish color giving the illusion of fires.

Indeed, there exists a famous climate/antropological theory based on a catastrophic volcanic winter episode refers to the Mount Toba eruption which is dated approximately 71,000 years ago. The Mount Toba eruption was apparently the second largest known explosive eruption over the last 450 million years and has been estimated to be forty times larger than Mount Tambora eruption (1815), which is considered the largest eruption known in human history. According to Stanley H. Ambrose of the University of Illinois at Urbana-Champaign [165] the Mount Toba colossal eruption caused an instant Ice Age with a sudden reduction of the summer temperatures by as much as 12 °C. This caused a widespread animal extinction and reduced the world's human population to 10,000 or even a mere 1,000 breeding pairs, creating a bottleneck in human evolution. The survivors from this global catastrophe would have found refuge only in isolated tropical regions, mainly in equatorial Africa from where modern humans are believed to descend.

Volcanic eruptions, for the most part, seem to influence climate on the time scale of a few years and do not explain secular trends. Evidently the above historical facts cannot be lightly dismissed, but it gives only a qualitative idea of how climate evolved during the last millennium. This qualitative pattern suggests that on a secular scale the solar variation made a significant contribution to climate change because solar activity experienced a maxi-

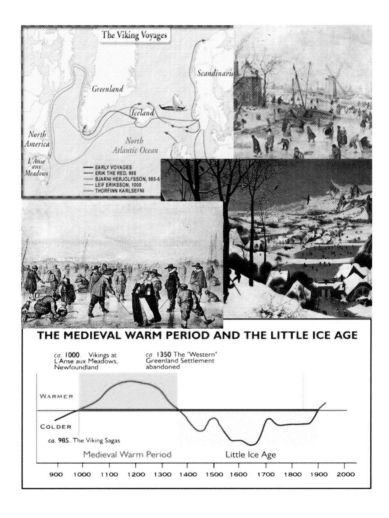

Figure 2.19: Global temperature change since the middle ages as inferred from historical records. The Vikings inhabiting Greenland from 1000 to 1350 has been commonly indicated as the most compelling evidence of a significant Medieval Warm Period. Several scenes, captured by the artists showing frozen rivers and lakes has been commonly indicated as evidence of a significant Little Ice Age. The figure shows a painting by Peter Bruegel the Elder (1525-1569), and two paintings by Hendrick Avercamp (1585-1634).

mum during the Medieval Warm Period and a minimum during the Little
Ice Age, as Figure 2.14 shows. However, qualitative inferences based on historical documents have a limited applicability in quantitative science. Long
data sets of global average temperature covering several centuries are, indeed,
required for an accurate interpretation of what happened.

In 1999 Mann and colleges [104] published the first paleoclimate Northern
Hemisphere (NH) temperature reconstructions covering the last 1000 years
making use of multiple temperature proxies such as tree rings, corals, ice core
historical records: see Figure 2.20A. This reconstruction is commonly known
as the *Hockey Stick* temperature graph[8] because it shows a characteristic
pattern with a slowly decreasing shaft from 1000 to 1900, and a sudden
increase, the blade, since 1900.

The *Hockey Stick* temperature graph has had a significant impact on the
global warming debate and has been widely published in the media. The
2001 United Nations Intergovernmental Panel on Climate Change (IPCC)
Third Assessment Report presented it as the most compelling evidence that
the warming observed during the last century has been anomalous and, therefore, could be attributed only to human activity. In fact, the *Hockey Stick*
temperature graph shows that the global temperature from the Medieval
Warm Period to the Little Ice Age decreased by just 0.2 °C degrees: a very
small change that has been interpreted as scientific evidence that climate is
quite insensitive to natural forcing and that the above medieval warm period and following cool period, if real, were at most regional, and not global.
Since 1900, instead the temperature warmed by about 1 °C, and only anthropogenic activity accounts for the difference. Subsequently, we discuss how
different models interpret this temperature reconstruction.

The Hockey Stick interpretation of the temperature data has recently
been seriously questioned because it is at odds with the new paleoclimate
temperature reconstructions proposed since 2002: see Figure 2.20B. Some
of these reconstructions, such as the one by Moberg *et al.* [113], present a
significant pre-industrial climate variation (approximately a 0.65 °C variation
from the Medieval Warm Period to the Little Ice Age) that can be explained
only if climate is very sensitive to natural solar forcing. Because solar activity
increased significantly since the Maunder solar minimum, during the peak of
the Little Ice Age, and in particular during the last century, the conclusion

---

[8]The name was coined by Jerry Mahlman, the head of National Oceanic and Atmospheric Administration's (NOAA) Geophysical Fluid Dynamics Laboratory.

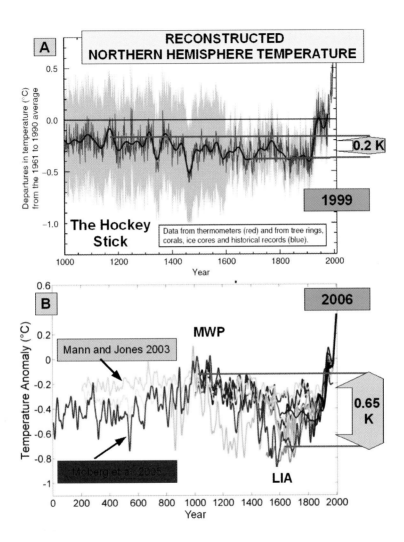

Figure 2.20: [A] The Hockey Stick temperature reconstruction by Mann *et al.* in 1999 [104]. [B] Comparison of several paleoclimate Northern Hemisphere temperature reconstructions available in 2006 [120]: The figure highlights two recent reconstructions by Mann and Jones [105] and Moberg *et al.* [113] (figures are redrawn).

of a predominant anthropogenic warming of the last century can no longer be considered as certain.

Right now, the significant uncertainty in the long temperature reconstructions and, as we have seen in the previous session, the significant uncertainty in the long solar activity reconstructions are responsible in no small measure to the heated controversy surrounding the global warming debate. The fact is that the climate models cannot be unambiguously tested against data that are still extremely uncertain and present large variability in secular trends. Indeed, the global warming debate is disrupted by its own theoretical and experimental uncertainties surrounding the climate network both its mechanisms and data.

Chapter 3

Information

We have now seen some of the ways the physics paradigm has been used historically to understand complex networks in the social, life and geophysical sciences. A common way to characterize the phenomena discussed was through the use of time series with apparently random fluctuations. The uncertainty associated with fluctuations occurs because networks in the social, life and geophysical sciences are very often more complex than those in basic physics. Continuing this train of thought in the present chapter, let us direct our attention to modeling data sets that do not lend themselves to the statistics of Gauss; the networks are more complex than that. The lack of a bell-shaped statistical distribution implies that the description of these networks have almost no historical parallels in physics and require examination using contemporary methods of analysis.

In this chapter we consider various measures of the complex dynamical networks discussed previously in terms of time series data and explore how these measures influence the design and understanding of networks. The uncertainty associated with these measures is a consequence of the randomness, irreversibility and unpredictability of what we can and what we cannot know about complex networks. The crux of the discussion is that information is a physical quantity that can be either bound or free. The amount of bound information determines the intrinsic complexity of a network, whereas the amount of free information determines the manner in which complex networks can interact with one another.

Islands of complexity develop with the passage of time and we examine a number of candidate measures that have been proposed to quantify the increased order in these islands in a surrounding sea of disorder. In this brief essay we do not attempt to investigate the mechanisms by which evolution favors increased complexity over time, this is too ambitious an undertaking for us here, but this would be a valuable area of research. Instead we discuss some of the problems and paradoxes that are entangled in the concept of complexity in the restricted domain of the physical sciences. We take this approach because, if complexity does have universal properties, these properties should be independent of the particular network being studied and therefore we choose the simplest context possible. For example, mathematical models of the generation and dissipation of fluctuations and their interactions with the data from complex networks are examined using the measures of entropy and information.

An issue related to the information paradigm of our understanding of nature is the principle of reductionism. This principle, in a nutshell, states that the process of understanding implies analyzing data for the purpose of arriving at generalizations. Such generalizations are very efficient descriptions of the world, reducing what we need to remember and enhancing our ability to communicate with one another. It is much simpler to communicate a natural law than it is to communicate the results of thousands of experiments upon which such a law is based. In strong form, reductionism states that to understand complex phenomena one needs only to understand the microscopic laws governing all the elements of the network that make up the phenomena. This reasoning implies that once we understand all the interactive parts of a process, we can add them up to understand the total. The whole is the sum of the parts; no more, no less. Such superposition is certainly adequate for understanding simple linear networks, but it is an incomplete description of the complex networks in the social, life and geophysical sciences.

To appreciate how the complications of complex phenomena have changed our view of the world we go back over half a century to the work on systems theory pioneered by Von Bartalanffy [13]. Generalized Systems Theory (GST) is a discipline describing networks of elements that very often organize themselves into patterns that cannot be understood in terms of the laws governing the individual elements. This self-organization constitutes the emergence of new properties that arise, for example, in living beings, which cannot be understood using reductionism alone, but require a more holistic perspective. This change in perspective, from the reductionistic to the holistic, in some ways parallels the passage from a deterministic to a probabilistic description; the meaning of information changes with the changing perspective.

Any science of networks must necessarily be multidisciplinary in that it should describe phenomena that cut across disciplinary boundaries. Consequently, we may argue that the schema we construct, guided by the paradigm of physics, is incomplete. On the other hand we believe that the information perspective gives an important advantage in addressing the difficult task of understanding complex networks. The GST approach to science has proven to be especially effective at the interfaces of well-established disciplines, for example, between biology and physics, the nexus of biology and chemistry, as well as the dovetailing of information and cybernetics, to name just a few.

Another reason to start at what has historically been the most fundamental level and use physics is by reviewing the arguments establishing that information is itself a physical quantity as was first articulated by Landauer [81]. He argued that information, like energy and entropy, is a property of the physical context in which the information is embedded. The majority of the physical science community overlooked this remarkable idea for nearly 30 years, but in the decade of the 90's it was recognized for the fundamental hypothesis that it is. We examine the proposition that information, like energy, is a physical quantity that can be separated into free and bound parts, and it is only the free information that can be used to measure the complexity of a network through its interaction with another network, say a measuring device.

The reader should not forget that the purpose of these excursions into historical reviews is to establish the context for the development of a science of networks. What ideas may contribute to that development are not always evident, making necessary the journey into the science of the nineteenth and early twentieth centuries to examine what was known empirically about networks and consequently what a science of networks must explain.

Information as an indicator of the structure within a time series is based on there being a probability measure for the process of interest. The probability density determines the entropy and therefore the information available about the underlying process. In a simple world the network of interest would develop according to the average of the time series with an uncertainty that grows with the standard deviation. It is remarkable that when the standard deviation for the complex network can diverge this Gaussian world view is no longer supported by data and must be replaced by the world view of Pareto. The Pareto distribution of income was discussed earlier, but in this chapter we concern ourselves with a variety of networks that have the Pareto statistical behavior including: economic, informational, neuronal, science, social and terrorist networks.

The Pareto distribution function implies certain interesting properties about the dynamics of complex networks. The complexity of a network is quantified by the power-law index, allowing the information gained about the intrinsic structure of science networks to be transferred to social, terrorist and other kinds of networks. The Pareto Principle is described, showing how neither rewards nor burdens are fairly distributed within a complex network. This imbalance is generic in the dynamics of the networks found in the world.

The notion of scaling encapsulates the structure of the statistical distribution and provides insight into the complexity of network dynamics. Of special interest is how one complex network perturbs the dynamics of a second complex network, resulting in the network of interest inheriting the statistics of the perturbing network. As the complexity of the perturbing network approaches that of the network of interest more and more information is transferred, until the complexity indices of the two networks are equal and the maximum amount of information is transferred.

The scaling property of fractals is ubiquitous in the world of the complex. Most, if not all, phenomena manifest scaling behavior once they have exceeded a certain threshold of complexity. Dynamic networks, economic networks, social networks, physical networks and language networks, all scale, at least in part. For example, scaling is observed in the number of sites with a given number of connections on the Internet. In fact, the connectedness of the communication/computer networks determines the security of the network against attack. Models have been constructed using different architectures for the connections to sites on a network. It was determined that those networks whose site connections scale, that is, the number of sites with a given number of connections are inverse power law, are the most robust against outside attack. The presence of inverse power laws, in physiological networks is indicative of evolutionary advantage [182]; it is feasible that the same is true in complex physical and social networks.

3.1   Entropy

Historically thermodynamics was the first scientific discipline to systematically investigate the order and randomness of complex physical phenomena, since it was here that the natural tendency of complicated physical processes to become disordered was first described. In this context the quantitative measure of *disorder* of proven value is entropy and thermodynamic equilibrium is the state of maximum entropy. Of course, since entropy is a measure of disorder, it is also a measure of order and complexity. If living matter is considered to be among the most complex of networks, for example the human brain, then it is useful to understand how the enigmatic state of being alive is related to entropy. Schrödinger [151] maintained that a living organism can only hold off the state of maximum entropy (death), by absorbing negative entropy, or negentropy, from the environment. This conjecture is central to understanding the relationship between entropy and information.

Entropy, like the length of a rod or the temperature of a piece of steaming apple pie fresh from the oven, is a physical quantity that is measurable. At the absolute zero of temperature the entropy of any piece of matter is zero. The French engineer Nicolas Léonard Sadi Carnot (1796-1832) in 1824 introduced the physical concept of entropy S into his study of heat engines. The amount of change in the entropy ΔS can be computed by calculating the ratio of the change in heat to the thermodynamic temperature. This concept of entropy requires the network to be macroscopic, isolated and in thermodynamic equilibrium. If any of these conditions is not fulfilled we cannot calculate the entropy exactly, but must be satisfied with the inequality $\Delta S \geq 0$, where the equality only applies for a reversible process over a thermodynamic cycle. The inequality implies that the change in entropy is positive for irreversible processes, that is, the entropy increases and the network tends towards disorder. The arrow of time is determined by the direction of increasing entropy for isolated networks and this in turn has been used to define what constitutes irreversibility.

Rudolf Julius Emanuel Clausius (1822-1888) rediscovered the *Second Law of Thermodynamics* in 1850 based on Carnot's work [24] and introduced the word entropy into the scientist's lexicon. According to the second law it is not possible to conduct an experiment in an isolated network whose only result is the spontaneous transfer of heat from a cold to a hot region. Heat cannot be transferred if the network is isolated, because work cannot be done on the network due to its isolation. Consequently the one-way flow of heat defines directionality for time. It then took Clausius another fifteen years to prove the mathematical form for the change in entropy postulated by Carnot. The Clausius definition of entropy is that used in thermodynamics today and does not rely on statistical concepts. However in a modern interpretation of order and disorder, fluctuations and statistics certainly play a role. The in-

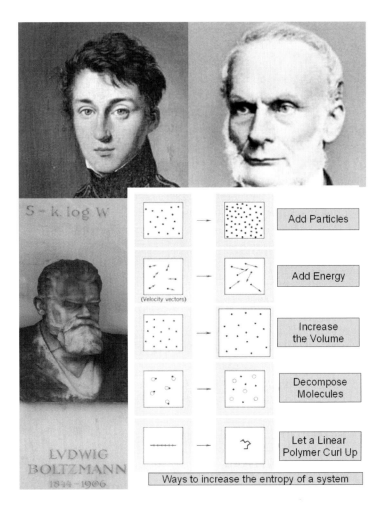

Figure 3.1: Nicolas Léonard Sadi Carnot (1796-1832) (top left); Rudolf Julius Emanuel Clausius (1822-1888) (top right); The grave of Ludwig Boltzmann (1844-1906) with the famous equation of entropy: $S = k \log W$. The diagram shows several ways the entropy of a network may increase.

vestigations of Ludwig Boltzmann (1844-1906) lead him, following Planck, to introduce entropy in the form now engraved where he can see it for all eternity, on the stone above his grave, $S = k \log W$. Here k is a constant that has the appropriate dimensions for entropy and has come to be called Boltzmann's constant; the equation is called the Boltzmann's relation. Imagine that the phase space for an isolated system can be partitioned into a large number of identical cells so that the probability of a particle occupying any of the cells in phase space is equally likely. The probability measure is the volume of phase space W consistent with the total energy of the network. The greater the volume of microstates, the more space available for the particles to rearrange themselves and consequently the more disorder that is possible.

The second law of thermodynamics is so well grounded in experiment that it provides a hurdle that every possible definition of entropy must overcome. Thus, we know that whatever definition we choose, entropy must increase or remain constant in a closed system, or more precisely it must be a non-decreasing function of time for a closed system in the limit where the network is described by an infinite number of variables. This regression to equilibrium, where, as we mentioned, equilibrium is the most disordered configuration of the network, is the irreversibility property of entropy. The future is therefore the direction in time that decreases the amount of order and leads towards the *heat death* of the universe. But this disordering does not occur equally at all places nor at all times; not all systems deteriorate in the short term, if they did then life would not be possible.

Erwin Rudolf Josef Alexander Schrödinger (1887-1961) [151] identified negentropy as that quantity a living organism obtains from the environment in order to keep from dissipating. Negentropy is the *stuff* that enables the organism to maintain its order. One way to quantify the local increase in the entropy of a network was developed by Ilya Prigogine (1917-2003). He (Prigogine) was able to develop and generalize this concept through the formation of dissipative structures, that are maintained by a flux of material and/or energy through the network of interest. Explicit in his ideas is that order is maintained by means of the network interacting with the environment, which implies that the network is not closed, in contrast to the assumptions of Clausius and Boltzmann. The dissipative structures of Prigogine are maintained by fluctuations which provide a source of energy for the network and dissipation which extracts energy from the network; both the fluctuations and dissipation arise from the interaction of the network with the environment. This balancing of fluctuations and dissipation maintains the flux through the system, which in turn supports the organization of the dissipative structure.

Implicit in the concept of entropy is the idea of uncertainty. Uncertainty means that not all the information one needs for a complete description of the behavior of a network is available. However, we can safely say that entropy is a measure of uncertainty, and like uncertainty, entropy is a non-decreasing function of the amount of information available to an observer. This connection between information and thermodynamics is quite important, and at this stage of our discussion we can say that it is the uncertainty that allow us to describe dynamical systems in thermodynamic terms.

Chaos, as we mentioned, is defined as a sensitive dependence on initial conditions of the solutions to a nonlinear network's equations of motion. Suppose we measure the number of distinguishable trajectories emerging from the numerical solutions to the dynamics of a nonlinear network. Any two initial points, say the starting points of two trajectories, may be located so closely that we are unable to discriminate between them. With the passage of time, however, the trajectories exponentially separate from one another amplifying the small difference in the initial states. This continuous separation is what makes them increasingly distinguishable. Thus, information about the network, in the sense of the visibility of an increasing number of trajectories, is generated by the dynamics. The number of trajectories, made visible by separating orbits, increases exponentially with time. The rate of exponentiation, which characterizes the mean rate of generation of this information, is the Kolmogorov-Sinai (KS) entropy [119]. The extreme of a zero growth rate implies that the underlying trajectories are regular and stable so no new information is generated as the dynamics unfold. The opposite extreme of an infinite growth rate implies that the motion is purely random, which is to say, all the trajectories are unstable and we are overwhelmed with information with the passage of time. Finally, new information in generated exponentially in time in the generic situation due to chaotic motion, in which case the KS-entropy is finite and determines the rate of generation of new information.

The entropy at time t, will be the conditional information entropy at time t, given the informational entropy at the previous point in time $t - \tau$. This connecting of the information at different time points, using the network dynamics, is continued until the probability is traced back to the initial conditions. Consequently, networks that are able to reach equilibrium can be said to *generate information*. It was argued by Shaw [154] that this is exactly the nature of chaos, the mechanism by which nonlinear dynamical networks generate information.

The physical theories that describe the dynamics of microscopic and macroscopic networks conserve probability in the sense that once a probabilistic point of view is adopted the total probability remains unchanged in time so that if the probability at some point in phase space increases, somewhere else it must decrease. In classical mechanics if one selects a volume in the phase space, and uses each point in this volume as an initial condition for the dynamical equations of motion, then the evolution of the volume may change its geometry, but not the magnitude of the volume itself. Equilibrium, however, means that the volume does not even change geometry. To achieve equilibrium the volume of initial states develops twisted whorls and long slender tendrils that can become dense in the available phase space [12]. Like the swirls of cream in your morning coffee, this property is called mixing. Not all dynamical networks are mixing, but this property is required to reach equilibrium starting from a non-equilibrium initial condition.

3.1.1 Wiener-Shannon Information

Historically, the mathematical and physical scientists that laid the groundwork for modern day statistics very often made their livelihood not by teaching at a University, but by calculating the odds in games of dice and cards and by casting horoscopes. Although the modern theory of probability had begun with the unpublished correspondence between Blaise Pascal (1623-1662) and Pierre de Fermat (1601-1665) in 1654 and the treatise *On Ratiocination in Dice Games* (1657) by Christiaan Huygens of Holland (1629-1695), the classical treatise on the theory of probability was written in the 1718, *The Doctrine of Chances*, by Abraham de Moivre (1667-1754) [114]. It is interesting that de Moivre's development of statistics and probability was directly coupled to his desire to predict. It seems that his book was highly prized by gamblers and that De Moivre was even able to correctly predicted the day of his own death by noting that he was sleeping 15 minutes longer each day and that he would die on the day he would sleep for 24 hours!

One may consider the context mundane, but gambling clarifies the position of the scientist. The scientist is the person that amidst the chaos of the world clings to the belief that there is an underlying order and a certain predictability; the scientist's goal is to enunciate the rules for this predictability. For example, de Moivre defined the average uncertainty in the outcome of many repetitions of a game as $I = -\langle \log P \rangle$, with the mean or average over the number of repetitions being denoted by the brackets, and P is the probability. This expression for I is determined below to be related to one definition of information. Of course, the concept of entropy, much less negentropy or information, did not exist in the lifetime of de Moivre, but his identification of the average uncertainty as a measure of what can be known about a random process does indicate the quality of his scientific/mathematical intuition. Note that the adoption of the logarithm is used to provide the additivity characteristic for uncertainty: the logarithm of a product is equal to the sum of the logarithms ($\log(ab) = \log(a) + \log(b)$). This property is important because the probability of a compound event composed of the intersection of statistically independent events is the product of the probabilities of its components, as was first stated in de Moivre's *Doctrine*.

In his 1948 book *Cybernetics* [189], the mathematician Norbert Wiener (1894-1964) was concerned with the problem of communication between humans, between machines, and between the two. In addressing this problem he makes the following observations:

> We had to develop a statistical theory of the amount of information, in which the unit amount of information was that transmitted as a single decision between equally probable alternatives. This idea occurred at about the same time to several writers, among them Dr. Shannon. The notion of the amount of information attaches itself very naturally to the classical notion in statistical mechanics: that of entropy. Just as the amount of information in a system is a measure of its degree of organization, so the entropy of a system is a measure of the degree of disorganization and the one is simply the negative of the other.

Consequently, Wiener expressed the information H by the equation $H = -\Delta S$ and because he was interested in the problem of noise and messages in electrical filters he treated entropy as a continuous variable in his analysis. From thermodynamics we know that the tendency of an isolated dynamical network is to deteriorate into disorganization and this is measured by entropy. The same observation can be made regarding the transformation of a message through one or more networks. Each such transformation degrades the order in the message and thereby information is lost, since information is the measure of order.

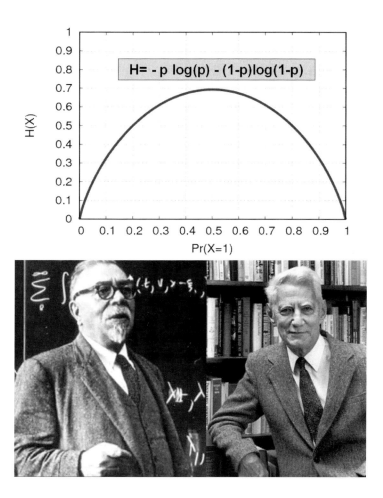

Figure 3.2: Norbert Wiener (1894-1964) (left) and Claude Elwood Shannon (1916-2001) (right). Top: information entropy of a coin toss as a function of the probability of it coming up heads (X=1). The entropy of the unknown result of the next toss of the coin is maximized if the coin is fair (that is, if heads and tails both have equal probability 0.5). This corresponds to the situation of maximum uncertainty. However, if the coin is not fair, there is less uncertainty and the entropy is lower. The entropy become equal to zero if the result is certain.

Claude Elwood Shannon (1916-2001) [153], on the other hand, was interested in the problem of coding information. He determined how to construct a formal measure of the amount of useable information a network contained and the problems associated with the transmission of messages within a network and between networks. In the previous section we introduced the idea that chaotic dynamical equations generate information. A less exotic version of this idea was developed by Shannon in his considerations of a discrete information source. He constructed a measure of the rate at which information is produced. He considered the situation in which there are n possible events with corresponding probabilities $p_1, p_2, ..., p_n$, but this is all that is known about which event will occur. The argument Shannon formulated was to determine how uncertain we are about the outcome; exactly what de Moivre wanted to know about his game of chance. Consequently Shannon obtained the same function as had de Moivre some two hundred years earlier[1], but with the science that had developed in the intervening years this function could be identified with entropy and subsequently with information.

Information can be expressed in terms of bits, or the number of binary digits in a sequence, and Shannon was able to prove that a network can be described by a function H that attains its maximum value when each of the possible states of the network have the same probability of occurrence. Notice that this is the same argument used in the Boltzmann relation; the assumption of maximal randomness (maximum uncertainty) has the probability being given by $p_i = 1/n$ and the entropy is given by $S = \log(n)$. Khinchine [72] did the mathematical analysis necessary to make Shannon's choice of information entropy mathematically rigorous.

3.1.2   The Physicality of Information

James Clerk Maxwell (1831-1879), considered by many to be the leading scientist of the nineteenth century, recognized that the kinetic theory of gases, which he invented and along with Lugwig Boltzmann refined, could in princi-

---

[1]Consider the probability in de Moiver's average uncertainty to be that for the outcome of a large number of games N, where the probability for the successful outcome of the jth game is p_j and the number of such outcomes is $N_j = Np_j$ for each j. Consequently the total probability is given by $P = p_1^{N_1} p_2^{N_2} \cdots p_N^{N_N} = \prod_{j=1}^{N} p_j^{Np_j}$. In this way de Moiver's average expression can be written as $I = -\frac{\log P}{N} = -p_1 \log p_1 - ... - p_N \log p_N = -\sum_{j=1}^{N} p_j \log(p_j)$, which is the definition for the Wiener-Shannon entropy, given a discrete number of outcomes.

ple, lead to a violation of the second law using information about the motion of molecules to extract kinetic energy to do useful work. In his 1871 book, *Theory of Heat*, he introduces his now famous demon:

> ... if we conceive of a being whose faculties are so sharpened that he can follow every molecule in its course, such a being, whose attributes are as essentially finite as our own, would be able to do what is impossible to us. For we have seen that molecules in a vessel full of air at uniform temperature are moving with velocities by no means uniform, though the mean velocity of any great number of them, arbitrarily selected, is almost exactly uniform. Now let us suppose that such a vessel is divided into two portions, A and B, by a division in which there is a small hole, and that a being, who can see the individual molecules, opens and closes this hole, so as to allow only the swifter molecules to pass from A to B, and only the slower molecules to pass from B to A. He will thus, without expenditure of work, raise the temperature of B and lower that of A, in contradiction to the second law of thermodynamics.

Léon Nicolas Brillouin (1889-1969), in his remarkable book *Science and Information Theory* [14], revealed that the paradox of Maxwell's demon is at the nexus of information and physical entropy. He reviewed the many resolutions to the paradox that have been proposed over the nearly hundred years between the publications of the two books. He pointed out that Leó Szilárd (1889-1969) [156] was the first to explain that the demon acts using information on the detailed motion of the molecules, and actually changes this information into neg-entropy, that is, information from the environment is used to reduce the entropy of the network. Brillouin, himself, resolved the paradox using a photon of light, against a background of blackbody radiation, which the demon must absorb to see a molecule of gas. We do not present Brillouin's discussion here, but we note his observation:

> ...every physical measurement requires a corresponding entropy increase.

He subsequently concludes that the average entropy increase is always larger than the average amount of information obtained in any measurement.

Weiner, in his book that started the field of cybernetics, suggested that the demon must have information about the molecules in order to know which to pass through the hole and at what times. He acknowledges that this information is lost once the particle passes through the hole and puts information and entropy on an equal footing by observing that the demon's acquisition of information opens up the network. The demon-gas network has increasing total entropy, consisting as it does of the sum of the entropy of the gas, which is decreasing, and the neg-entropy (information), which is increasing. Ball [7] notes that Szilárd had previously concluded that:

> ...the second law would not be violated by the demon if a suitable value of entropy were attributed to the information which the demon used in order to reduce the entropy of the rest of the system.

The key feature in the resolution of Maxwell's demon paradox is the existence of dissipation, that is, the erasure of memory, in the information cycle. This occurs in Brillouin's argument through the requirement that the photon energy exceed that of the blackbody radiation and in Wiener's discussion, in the observation that the particle forgets its origin once it passes through the hole. Landauer [81] indicates that these early arguments can be summarized in the statement:

> ...the erasure of the actual measurement information incurs enough dissipation to save the second law.

independent of any specific physical mechanism. Bennett [10] was able to pull the threads of all the various arguments together, in the context of doing reversible computation, and in so doing obtain what is considered to be the final resolution of the Maxwell demon paradox.

This final resolution of the paradox involves a concept that is unfamiliar to most physical scientists and is called *Landauer's Erasure Principle*. In essence the principle states that memory cannot be erased without the generation of heat, in fact without an energy cost of $W = kT \log(2)$ per bit of information. This energy cost is a direct consequence of the physical nature of information. Correspondingly, for every bit of information erased, there is a minimum increase in entropy of the system $\Delta S = k \log(2)$. The cost in energy and change in entropy come about because erasure must be done by performing a physical operation.

Consider two thermally isolated boxes, labeled A and B, each with a partition in the middle. In A there are n gas particles to the left of the partition and in B there are n gas particles to the right of the partition. If we associated 0 with the left side of a box and 1 with the right side; then one box has all the particles in state 0 and the other has all the particles in state 1 and the configurations are distinct. With the removal of the partitions the gas particles begin to diffuse outward due to internal collisions. After a relatively short period of time the particles no longer remember whether they began on the right in B state 1, or on the left in A state 0. Each particle has a 50-50 chance of being in the state 0 or 1, so the change in entropy per particle is $k \log(2)$ and the total change in entropy in each box is $\Delta S = nk \log(2)$.

Now we use a piston in each box to slowly push all the particles to the left-hand side, that is, into state 0. Once this is done a partition is reinserted in both boxes A and B. Looking at the two boxes it is no longer possible to distinguish one from the other since the particles in both boxes are in state 0; consequently we have erased the memory of the initial state. However we know that it took an entropy change equal to $nk \log(2)$ to change the configuration from one in which the particles in the two boxes are in different states to one in which they are in the same state. Therefore there is an entropy increase associated with the erasure of the memory of the initial state.

Here we focus our attention on Landauer's interpretation of information as a physical phenomenon. Rather than being the abstract quantity that forms the basis of reified mathematical discussion in texts on information theory, which it certainly does, information itself is always tied to a physical representation. Whether it is spin, charge, a pattern of human origin, or a configuration of nature's design, information is always tied to a physical process of one kind or another. Consequently, information is physical in the same sense that entropy and energy are physical. Landauer explores the consequence of this assertion by noting that the physical nature of information makes both mathematics and computer science a part of physics. We shall not dwell on this prioritizing of the sciences, but shall review some of the consequences of this perspective.

Let us divide the world into the macroscopic and the microscopic and assume that the statistical entropy can describe the microscopic network. Let us follow the discussion of Ebeling [44] and assume that the macroscopic world is described by an order parameter that can take on M values, each with a probability p_j. This discrete representation of the macroscopic world is different from the Hamiltonian form of the dynamics so often used in the physics literature, but using Shannon's formula allows us to write an entropy

associated with each node of the macroscopic network. The bound entropy
is determined by the microscopic degrees of freedom of the network and is
not available as useable information. In analogy with free energy, which is
the energy available to do work, we have free entropy, which is the entropy
available to produce useful information. Thus, the free entropy is precisely
the neg-entropy discussed by Szilárd, Wiener, Brillouin and so many others.
Ebeling explains that free information corresponds to the macroscopic order
parameter entropy and may be changed into thermodynamic entropy which
is the entropy generated by and bound to microscopic motion.[2]

The separation of the total entropy into a piece depending on the network
of interest and a piece depending on the conditional entropy of a subnetwork
and the correlation with the network of interest was constructed by Khinchine using a discrete representation. In the discrete network the conditional
entropy of the subnetwork was proven to be less than or equal to the entropy
of the subnetwork when it is independent of the network of interest. Thus,
it is clear that the macroscopic network suppresses the disorder of the microscopic subnetwork at finite times. The entropy of the network of interest
determines the information obtained with a measurement of the macroscopic
variables at time t and the entropy of the microscopic network provides the
amount of additional information provided by a subsequent realization of the
ambient network. When the time scales of the microscopic and macroscopic
networks are well separated the entropy of the former network is maximum
and consequently the microscopic network is in thermodynamic equilibrium.

Natural messages are often sequences of symbols on an infinite alphabet
due to the continuous character of the network generating the signal. However, the information that is generated by such networks is often discrete
rather than continuous. For example speech is continuous in frequency and

---

[2]Let us describe the microscopic world in terms of the set of variables Γ and $P(\Gamma)d\Gamma$
is the probability that the microscopic world is in the interval $(\Gamma, \Gamma + d\Gamma)$. If the microscopic world is coupled to a network of M macroscopic states then $P(\Gamma) = \sum_{j=1}^{M} p_j P_j(\Gamma)$
and $P_j(\Gamma)$ is the conditional probability density that the microscopic states are in the
indicated intervals, conditional on the order parameter being in the macroscopic state
j and is normalized. Using the continuum form results in separation of the entropy
into two parts $S = S_B + S_F$ where the 'free' entropy is given by the discrete expression $S_F = -\sum_{j=1}^{M} p_j \log(p_j)$ and the 'bound' entropy is given in terms of the entropy of the microscopic network given a macroscopic state $S_B = \sum_{j=1}^{M} p_j S_j$ with
$S_j = -\int P_j(\Gamma) \log P_j(\Gamma) d\Gamma$. The free entropy is seen to add to the entropy determined by
the distribution of macroscopic states.

amplitude, however, the alphabet from which words are constructed is finite and the messages constructed from them are discrete. Ebeling and Nicolis [43] discuss the structure of individual symbolic sequences and the statistical properties of large assemblies of such sequences, with special emphasis on the role of correlations in the probability density of a given sequence generated by a dynamic process.

The message of interest in the context of communication theory is expressed in terms of an alphabet of alternatives; more generally the message is embedded in the data discussed in the previous chapter, which as we saw can be either continuous or discrete. In either case the degree of order in the data constitutes the information in which we are interested. The definition of information developed above can be used to assist us in gaining insight into a given data set. The probability distribution with the same mean and variance as the given data, but with minimum information otherwise, is the distribution of Gauss. What this means is that if the statistics of the data are described by the Gaussian distribution then the only information that can be obtained from the data is the mean and variance. If there is any additional order in the data set, then the resulting distribution would have a different form and that is what is found almost universally outside the physical sciences. This takes the discussion back to Pareto and his distribution.

3.2 Pareto's Law

So now let us try and put a face on Pareto[3] in order to better understand how he envisioned society and its workings. The setting is a typical European classroom of the nineteenth century. From the front of the classroom one can look up into the dozen-or-so tiered-rows of wooden benches and writing surfaces and see the last student in the last row, however low s/he slides down into the seat. The professor is well-dressed, in a suit, waistcoat and tie, which emphasize his diminutive physique. His beard is trimmed short, as is his dark hair, and to all appearances, aside from his bearing; he is a most unremarkable man, that is, until he begins to speak. When he speaks his eyes burn brightly and the cadence of his words and their logic capture the audience scattered around the room. It is not just his words that pull adolescent listeners out of their lethargy; while waiving his arms in mid-sentence, he goes to the blackboard and scrawls a diagram. Indicating one

---

[3]All the quotes in this subsection are taken from Pareto's two volume work. [124]

point in the upper left of the sketch with an A and another at the lower
right with a B, and without interruption to his monologue he strikes a line
connecting A to B, underscoring the necessity of their connection.

As he speaks it becomes clear that this is a man of conviction; a man
who has been targeted by the authorities in Italy as a troublemaker. Before
coming to Switzerland, he was frequently trailed by police and intimidated
by hired thugs, but being well trained with the sword, a crack shot with a
pistol and equipped with an aristocratic sense of honor, the Marque never
let himself be physically intimidated.

In the classroom his words are passionate, carrying his belief in what he
said, as well as his desire to convince his students of their truth:

> It is an indisputable fact: human beings are not equal physically,
> intellectually or morally.

Anticipating a negative reaction from the traditionally liberal student
body he looks around the room, searching for descent. He wants to frame his
argument to soften the scowl of the skeptic or dispel the unconcern of those
that are bored:

> The curve of the distribution of wealth in western societies varies
> very little from one period to another. It probably relates to the
> distribution of the physiological and psychological characteristics of
> human beings.

He knows that his explanations lack the concreteness of the data from
which he drew his conclusions, but his words are consistent with his more
general theory of social order, something his students will learn if they stay
the course:

> Supposing men to be disposed by strata according to other charac-
> teristics, for example, by intelligence, by aptitude for mathematics,
> by musical, poetic and literary talents, by moral characteristics and
> so on, we shall probably get curves whose forms are more or less
> similar to that which we have just found by ordering men according
> to the distribution of wealth. This wealth-curve results from a large
> number of characteristics, good or bad as it may be, which taken to-
> gether are favorable to the success of the individual who seeks wealth
> or, having acquired it, conserves it.

Pareto did not know with certainty how prophetic his remarks would prove to be, but the empirical evidence concerning the truth of his observations and speculations have resisted the attacks of scientists over time.

The professor knew from experience that presenting ideas in mathematical form, no matter how simple, would not touch the hearts of students. They would not see that the concept is universal, that the distribution of income is not just true for this or that society, but the distribution is true for all societies, in a way that does not depend on any particular individuals. He could not communicate to them the import of his findings; after all, they were just beginning to study the nature of society, so he begins simply:

> Let us consider a community in which incomes are represented by the following graph.

Abruptly he turns to the blackboard, stretches out his arm and sketches a curve without saying a word. He then turns back to the class with a somber expression and explains, using the chalk as a pointer over his shoulder (see Figure 3.3):

His eyes dart around the large room, peering into the faces of the students, trying to determine what, if anything, they are absorbing from what he is telling them. Their response is passive; there is no glimmer of understanding. He explains more forcefully:

> ...the molecules composing the social aggregate are not at rest: some individuals are becoming rich, others poor. Appreciable movements therefore are occurring within the figure m-s-t-b-s-a-m. In this the social organism resembles a living organism.

This analogy of society with living processes seems to kindle their interest. He can see some students lean forward and others adjust their position to better see the blackboard and so he continues:

> The external form of a living organism — a horse, for example — remains almost constant, but internally extensive and varied movements take place. The circulation of the blood is putting particular molecules in rapid motion; the processes of assimilation and secretion are perpetually changing the molecules composing the tissues. The analogy goes even further.

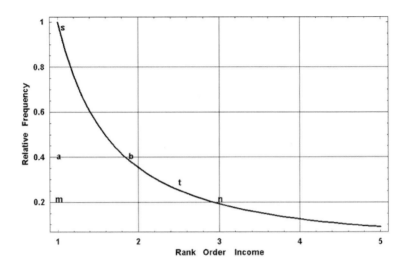

Figure 3.3: The community is represented by the surface m-n-t-b-s-a, individuals being the elements of this surface. We know that the form of the curve n-t-b-s varies only slowly. This form may be supposed almost constant on average (modified from Figure 4 in Pareto [124]).

He has their attention now, almost all the students were looking at him and not at one another or at the desk. They understood the actions of the body and by analogy they could see how the life's blood of society, that is, wealth, could be exchanged among society's members. Sex and money, the two things that could be counted on to get a young person's attention can now be brought together in the discussion. So he continues:

> It is only the form of the curve n-t-b-s which remains constant; the absolute dimensions of the figure may, and do in fact, change. Thus the foal becomes an adult horse. Similarity of form is maintained, but the size of the animal changes considerably.

How is he to convey the notion of constancy of form, the self-similarity he has just mentioned in passing? How can he capture the constancy of the empirical curve in spite of the dynamics of the social process being considered? He plunges ahead:

...experience shows that the form of the curve only changes with great slowness; in fact, remaining well-nigh constant. We therefore conclude from this that each social stratum receives almost as many individuals as it loses. In other words: there are, for example, almost as many individuals every year entering the class of those who have an income of 4,000 to 5,000 francs as there are individuals leaving this class.

This idea of having a continual flux of people through various income levels, but leaving the relative number of people in each income level constant is not an easy concept to grasp. He can see that some students are having difficulty with it and he immediately goes on to explain:

It must be clearly understood, as we have already strongly emphasized, that what is involved here is only an average phenomenon. It would be utterly absurd to claim that the number of persons in a month, or even in a year, who join the class of people with an income of 4,000 to 4,010 francs, or even of 4,000 to 5,000, is exactly equal to the number leaving these income groups.

Now he is at the core of his findings. That the result of years of data analysis could be distilled into one equation still leaves him with a feeling of awe, but he is convinced that:

In all places and at all times the distribution of income in a stable economy, when the origin of measurement is at a sufficiently high income level, will be given approximately by the empirical formula $y = ax^{-\nu}$ where y is the number of people having income x or greater and ν is approximately 1.5.

He smiles as he reveals this scientific solution to one of the great mysteries of society. This is a universal law of man's interaction with man and harbors one of the world's great truths. He hopes to be able to convey the significance of his result without overwhelming the audience:

These result are very remarkable. The form of this curve seems to depend only tenuously upon different economic conditions of the countries considered, since the effects are very nearly the same for the countries whose economic conditions are as different as those of England, of Ireland, of Germany, of the Italian cities, and even of Peru.

But now he has to wrap it up. He has certainly given them something tangible to take with them and think about. He decides to leave the discussion of the mechanisms giving rise to the distribution of wealth to another day:

> The causes, whatever they may be, which determine the distribution of ownership of income may operate with variable intensity. An extreme case would be that in which each individual found himself infallibly placed in that income level which corresponded to his capacities. Another extreme case would be one in which each income level was closed caste, no exchange taking place between one level and another. The cases occurring in reality are intermediary between these two extremes.

Pausing for a moment to see if he has stirred their curiosity Pareto exits the front of the room from a side door. While closing the door he looks back at the students who are getting up to leave the lecture hall. He wonders, not for the first time, if he had been so docile when he was in their place. These are new and exciting concepts he is presenting and yet most of the time his students fail to react to what he is saying. He shrugs his shoulders, closes the door and thinks about how he should organize the ideas for his next lecture.

In itself, constructing a distribution of the inverse power-law type would not necessarily be particularly important. Part of what makes the Pareto (inverse power-law) distribution so significant are the sociological implications that Pareto and subsequent generations of scientists were able to draw from it. For example, he identified a phenomenon that later came to be called the *Pareto Principle*, that being that 20% of the people owned 80% of the wealth in western countries. It actually turns out that fewer than 20% of the population own more than 80% of the wealth, and this imbalance between the two groups is determined by the Pareto index. The actual numerical value of the partitioning is not important for the present discussion; what is important is that the imbalance exists.

In any event, the 80/20 rule has been determined to have application in all manner of social phenomena in which the few (20%) are vital and the many (80%) are replaceable. The phrase "vital few and trivial many" was coined by Joseph M. Juran in the late 1940s and he is the person that invented the name Pareto Principle and attributed the mechanism to Pareto [70]. The 80/20 rule caught the attention of project managers and other corporate administrators who now recognize that 20% of the people involved in any given project produce 80% of all the results; that 80% of all the interruptions come from the same 20% of the people; resolving 20% of the issues can solve 80% of the problems; that 20% of one's results require 80% of one's effort; and on and on and on. Much of this is recorded in Richard Koch's book *The 80/20 Principle* [75]. This principle is a consequence of the inverse power-law nature of social phenomena.

3.2.1 Economic networks

Scientists take pride in acknowledging colleagues that have made what they consider to be fundamental contributions to human knowledge. The form such acknowledgment often takes is naming something after them. Experimentalists most often have their name given to a device, probably one that they invented. For example, the Wilson Cloud Chamber records the spatial trajectories of high-energy particles and is named after the Noble Prize winning physicist who invented it. Theoreticians most often have their names associated with laws; laws that encapsulate a large number of experiments into a single theory. Newton's laws of motion come to mind. His laws of motion provide the context for classical mechanics through the concepts of inertia (mass) and force (acceleration); concepts empirically developed by Galileo.

However, there exists a third class of acknowledgment, a category that captures experimental results, but does not necessarily provide a theoretical explanation for those results. For example, Boyle observed after a large number of experiments that as one decreases the volume of a gas the pressure increases, such that the product of pressure and volume remains constant. The empirical constancy of this product is known as Boyle's Law. We should also point out that Boyle's Law is only valid if the temperature of the gas does not change during the experiment; if the temperature does change then one obtains the Perfect Gas Law in which the product of pressure and volume of the gas is not constant but is proportional to the temperature of the gas. This last example shows that all physical laws have regions of validity; none of them are true under all conditions. So when a law is discussed, part of that discussion concerns its domain of validity and applicability. The distribution of income discussed by Pareto is no exception.

The empirical distribution of income has the inverse power-law form given by Pareto in his lecture of the last section. Assuming that the mathematical equation for the inverse power-law distribution is a mystery to you, let us see if we can find a way to reveal what is contained in that masterfully terse description of this social phenomenon, that being, how income is distributed among the members of a society. As with most such discussion it is usually best to begin with data. In Figure 3.4 we display income data for the years 1914 to 1933 inclusive, for the United States.

Note that in Figure 3.4 the data are displayed on what is called log-log graph paper, that is, the logarithms of the variables are recorded rather than the variables themselves. It can be seen that on such graph paper each factor of ten in population is one division and each factor of ten in income is also one division. Therefore 1 is graphed as 0; 10 is graphed as 1; 100 is graphed as 2; 1,000 is graphed as 3 and so on. Those more mathematically minded will see that each of the original numbers is ten raised to a power and it is this power that is being graphed rather than the number itself, that is what we mean by the logarithm. In terms of logarithms we can see that the income distribution is a sequence of straight lines with negative slopes. The slope of the distribution is the parameter that Pareto labeled with the Greek letter nu (ν) and which is often referred to as the inverse power-law index. The distributions of income in the United States are remarkably similar from 1914 to 1929, being a sequence of parallel line segments. The end of the curve where the date is recorded corresponds to the number of incomes in excess of one million dollars in that year. The distributions for the various years are separated in order to better compare them visually.

We can see an almost steady increase in the location of the end point of the income distribution curve, starting in 1921 and ending with the year of the Great Stock Market Crash of 1929 which heralded the Great Depression. This vertical increase in the location of the end point from one year to the next of the distribution indicates that more and more people were making in excess of one million dollars as the years went by. It was the *roaring twenties* and economic times were fabulous; then came the stock market crash of 1929. In 1930 the end point of the distribution curve plummeted and the fraction of society living the good life decreased by nearly a factor of ten. For every hundred people on the gravy train before the crash, there was only one after the stock market stabilized in 1932. It is interesting that throughout this season of prosperity the number of poor seemed to be constant, at least in terms of income level. But perhaps the number of people willing to pay income taxes at this lowest level of society is an invariant.

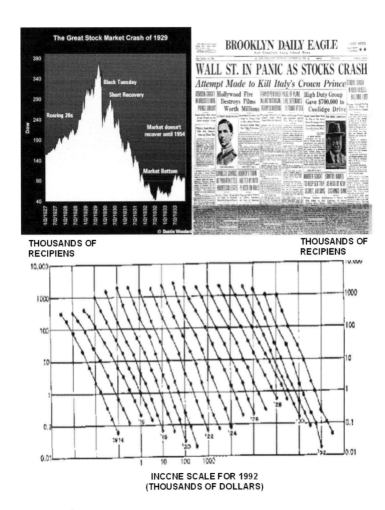

Figure 3.4: Comparison of income distribution in the United States from 1914 to 1933, inclusive [38]. Both the scale of the recipient population and the scale of the income are the logarithm of the appropriate variable. The vertical lines are one cycle (factor of ten) apart, as are the horizontal lines, the scale shifting one-half cycle to the right for each successive year of data. The point nearest the date in each case measures the number of incomes in excess of $1,000,000 in that year [115].

The careful observer may also notice that the slope of the income distribution curve was approximately constant before, but changed after, the crash. The 1930 distribution is steeper than the 1929 distribution, indicating that the value of the power-law index is greater after the crash than it was before. In fact, on the day known as "Black Tuesday" (October 24) the people who owned stocks, saw them loose about one tenth of their value: the stock market suffered a loss of approximately 12 percent, which affected the entire economy for the following 2-3 years. In such a scenario the Pareto index increases as a result of the top income class being expected to have the greatest loss from a decrease of the value of the stocks because their income is significantly based on investment [141]. But by 1933 the slope had regained its pre-crash value. The slope parameter is apparently responsive to the aberrant internal dynamics of society occurring as a consequence of the crash. But this section is not intended to be a study of the history of the stock market, or even of economics, except in an incidental way.

Our interest is in the Pareto distribution itself and the properties of the phenomena that it was subsequently shown to describe. But to understand how this distribution of income arises let us attempt the kind of argument one may find in a typical academic setting. We recognize that in order to have a specified level of income a person must have a certain combination of skills. In today's society, for example, a person must have a college degree; at least a Bachelor's degree and in the world of business very often a Master's degree. In addition to this educational requirement an individual must be able to communicate well, at least giving the impression of an education even when the degrees are absent. Being able to speak to colleagues, customers and to the public is often a gait keeper to upper income levels. Those able to present a compelling argument while putting a person at ease and being empathetic can expect the keys to the executive washroom; while those that stumble and hesitate in their speech find themselves looking over the top of a cubicle in a Dilbert cartoon. Writing is also a limiting ability. Those that cannot express their ideas in print are usually sidetracked at some intermediate level of management, whereas those that can spread it thick and easy, chair the boardroom meetings. But communication is also about presentation. How one dresses, the tone of voice, the choice of words, and the pattern of gesturing; all contribute to what is being communicated. The passive aggressive that speaks in a barely audible whisper may force the listeners forward to hear what s/he is saying, but the concentration is often counter-balanced by the irritation the listener experiences while straining to hear. The overtly aggressive person, who stands so close that you back away, only to have them put their hand on your arm or shoulder to pull you back, while talking directly into your face, is equally offensive. Both these personality types, when this behavior is automatic, rather than being in control, are self-limiting in how far they can rise in the corporate structure. However such devices can also be used as tools to undercut the opposition and are very effective when used selectively.

These devices notwithstanding it is not just one's technical skills or personality that determines how high one can fly; it is also a matter of ambition, or as some may say, motivation or aggressiveness. It is not likely that one will achieve a high level of success that is unsought, except in the rarest of cases. A factor that is truer today than in previous generations is the importance of adaptation. Learn about a new product, explore a new market, put together a new work team, all these areas require an individual able to adapt to a rapidly changing business environment; often changing to a new paradigm before the old one has been properly mastered. For example, the rate of change of a business product may be related to Moore's law in the computer industry. Moore's law states that every eighteen months the amount of data that can be stored on a computer chip doubles. This may not seem like a particularly fast rate of growth, but what it means is that in fifteen years the data stored on a chip increased by a factor of one million. So a kilobyte of memory in 1980 became a gigabyte of memory in 2000; the floppy disk was pushed out by the CD, which in its turn is now yielding market share to the thumb drive.

We have listed a number of talents that may assist a person in elevating their level of income: education, communication, personality, charm, adaptability, ambition, empathy, productivity and work ethic. These may not even be the most important characteristics, but they are representative of the fact that making money is not in itself a simple task, but requires an array of talents. This was just as true in the nineteenth century when Pareto uncovered the distribution of income, as it is in the twenty-first century even though some elements on the list have dropped off and some new ones have been added. The distribution of income is dominated by the few out in the high income tail because making money is a complex process and leveraging the ability of others to enhance your own income is intrinsic to the process.

If the world were a fair place then each of us would have approximately equal shares of the above talents. We would all have an equivalent level of skill, like the graduates of a self-selective trade school, say, Harvard School of Law. Each of us upon graduation would have a blue blazer, gray slacks, white shirt and maroon tie to go along with our degree as we forage through available job interviews. The uniform of the recent graduate allows for some variation, but very few would flout social convention to the extent of going without a tie or wearing white socks to an interview. As recognized by Pareto the distribution of abilities is more like his inverse power law than like the bell curve of Gauss.

3.2.2   Science networks

The inverse power-law of Pareto has a number of remarkable properties, not the least of which is the fundamental imbalance manifest in the underlying process it describes. In Figure 2.5 the difference between the Gauss and Pareto distributions is made apparent. The bell-shaped curve is seen to be symmetric around the average value, with variations a given level above the average being of equal weight to those at an equivalent level below the average value. In the distribution of height, for example, symmetry means that the fraction of the population one inch taller than the average is equal to the fraction of the population that is one inch shorter than the average. At the university, symmetry translates into class grades; where the imposition of the bell-shaped curve guarantees the same number of A's and F's, the same but greater number of B's and D's, but the greatest number of students have a grade of C. Leaving aside all the specious arguments given over the years as to why this curve ought to be applied to the distribution of grades in college, we now know that the Gauss distribution only works if one assumes the archetype of the average man discussed earlier. If the true complexity of learning is taken into account during the evaluation process there is no reason why the distribution of Gauss should have anything to do with course grades. So let us examine a somewhat simpler network of learning, that of the scientist.

The Pareto distribution, with its long tail, is not symmetric and this lack of symmetry Pareto interpreted as a fundamental imbalance in society. Moreover, Pareto saw that this imbalance is predictable because most results in society arise from a minority of causes. In this section we examine some sociological areas strongly influenced by this imbalance.

In the social sciences the bell-shaped curve introduced the notion of a statistical measure of the performance of a process relative to some goal. An organization is said to have a quantifiable goal that can be measured, when a particular measurable outcome serves the purpose of the organization. The number of students advancing and eventually graduating serves the purpose of the university, the number of recruits successfully completing basic and advanced training serves the purpose of the military, and the number of sales serves the purpose of most manufacturing companies. A sequence of realizations of this measurable outcome produces a set of data; say the number of research articles published per month by members of a research laboratory. Suppose a laboratory's goal is to achieve a specified publication rate. This ideal becomes the standard against which the actual rate of publications is compared. However, the empirical publication rate is not a fixed quantity, but varies a great deal from month to month. If 68 percent of the time the lab essentially achieves its goal, the lab is said to be functioning at the 2-sigma level, where sigma is the standard deviation. The 2-sigma level is the level attained within the 'C' region of the distribution, that is, between +1 and −1 in the standard from of the Gauss distribution in Figure 2.1. Higher sigma levels indicate that greater percentages of the outcome meet or exceed the ideal rate. Consequently, if the Gauss distribution is a valid representation of the processes then sigma could be used as a measure of success with which to compare different organizations.

The distribution of Gauss works well in describing simple physical phenomena, and even in describing some manufacturing processes, but it does not correctly quantify measures for sociological networks of interacting individuals such as scientists and engineers. Consequently, collections of scientists and engineers cannot be measured in the same way that manufacturing widgets are measured. We advocate a view of scientific measurement that takes into account the nonlinear, multiplicative nature of science networks as opposed to the linear additive nature of manufacturing networks. We should emphasize that this argument is equally valid when applied to any social group with multiple internal interactions focused on achieving a complex set of goals, such as a university faculty or the members of the Officer Corp.

From Figure 2.5 we deduced that social phenomena are unfair in the real world. If things happened according to Gauss, then everyone would be similarly situated with respect to the average value of the quantity being measured. If the normalized variable were a measure of income, then everyone would make approximately the same amount of money. If, rather than money, the distribution described the number of scientific publications, then the world of Gauss would have most scientists publishing the same average number of papers with a few publishing more and a few publishing less than the average. So the scientific world would also be basically fair. In fact when measures of scientific productivity are discussed, say in the promotion of a faculty member at a research university or a staff scientist at a research laboratory, the average number of publications is treated as if the world were fair, resulting in a variety of arguments that have nothing to do with the measure of scientific achievement.

However, fairness is not the way of the world. In the real world, Pareto's world of the inverse power law, there may not even be an average value and people may have almost none of the variable or they may have a great deal of it. So using the example of income, we find people at the level of poverty and people making millions of dollars a year and all those in between. The number of publications is no exception in this regard, Lotka [98] determined that the distribution in the number of scientists having a given number of publications is, in fact, inverse power law, just like that of income: see Figure 3.5. The difference between the two distributions is the parameter determining the rate of decrease in the inverse power-law; the power-law index or the Pareto index.

As we have seen, phenomena described by inverse power-law distributions have a number of surprising properties, particularly when applied to society. In the case of publications, it is possible to say that most scientists publish fewer than the average number of papers. It is definitely not the case that half the scientists publish more than the average and half publish less than the average, as they would in the world according to Gauss. However, the vast majority of scientists publishes less than the average number of papers, say about 95%. For every 100 scientists who produce only a single paper in a certain period, say a year, there are 25 that produce two in that year, 11 that publish three, and so on [161]. Expressed somewhat differently, we could say that 1 in 5 scientists produce 5 papers or more; 1 in 10 scientists publish at least 10 papers and so on. Consequently, if the number of papers were the measure of quality used by management to evaluate a laboratory, most labs would rate well below average. However, this would be a distortion of the true quality of the science being done in the lab, since this application of the measure presupposes a Gaussian world view as the basis for comparison both among the members of the research group and in comparing different research laboratories.

In addition to the productivity measured by the number of publications, the value of a scientist's research is also gauged by the scientific links they establish, both in quality and quantity. One way scientists form professional

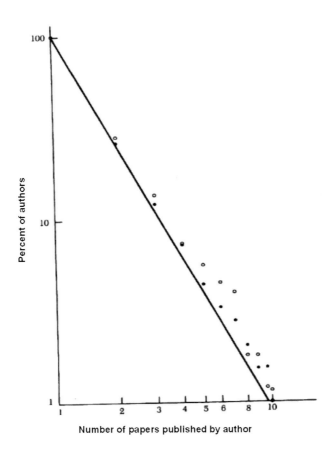

Figure 3.5: Lotka Law: The percentage of authors publishing exactly n papers, as a function of the number of papers. The open circles represent data taken from the first index volume of the abridged *Philosophical Transactions of the Royal Society of London* (17th and early 18th century), the filled circles those from the 1907-1916 decennial index of *Chemical Abstracts*. The straight line shows the exact inverse power-law of Lotka. All data are reduced to a basis of exactly 100 authors publishing a single paper [98].

links is by means of citations, that is, by referring to one another's published work. The more often a piece of research is cited by colleagues, the more valuable that research is thought to be to the scientific community as a whole. Consequently, the greater the number of citations accorded a given paper the more significant the underlying research is thought to be. We must recognize that the purpose of scholarship is to aggregate the talents and findings of many people. Nowhere is this more concentrated than in science where the focus is so intense that even a person of modest talent may, on occasion, make a significant contribution to the store of human knowledge. This is often apparent in the collaborations and interconnections scientists make in the course of their research. So we ought to examine the phenomenon of citation a bit more carefully, since it is, after all, another kind of science network. it makes visible what the seventeenth century natural philosopher Boyle called the invisible college.

First of all, even with the speed of modern communication, there is a time lag between publishing a paper and the content of that paper becoming known to the research community. The increased speed of communication is in large part offset by the multiple competitors for a scientist's attention. A few decades ago it may have taken a year or two for a paper to be refereed and published in a high quality journal, whereas today a prepublication version of the paper can be posted on a scientific website such as the Los Alamos National Laboratory, at the same time the manuscript is sent to a journal to be refereed. However, instead of the few dozen published papers in a journal one had to read once a month in that earlier time, the website has that many manuscripts posted per day that vie for attention. Thus, the research must be advertised in colloquia, conferences, workshops, emails and by any other means that stimulates colleagues to attend to the research. Thus, today a delay time of perhaps three years would be realistic to allow the research to be adequately disseminated. Using this estimate we reach the paradoxical conclusion that the increased speed of posting information has actually decreased the rate at which information spreads through the scientific community. We emphasize that this is a conjecture waiting to be disproved.

Here again the empirical result is that the number of citations to published research papers is an inverse power-law distribution. The number of citations obtained by papers published in a given year is depicted in Figure 3.6. Here is where the strangeness of the inverse power-law distribution becomes even more apparent. Suppose a faculty member is being considered for tenure and promotion from Assistant Professor to tenured Associate Professor. A senior member of the Promotion and Tenure Committee argues that the nominee's publications have only an average number of citations. He further emphasizes that the university only wants outstanding tenured faculty, not individuals that are merely average. How can tenure be granted to a person who does not excel? This apparently reasonable argument has all too often been sympathetically received by such committees in the past, resulting in first class scientists being denied tenured at the 'best' institutions and continuing their careers elsewhere.

The fallacy in the above tenure argument can be found by looking at the citation distribution more closely. First of all we see from Figure 3.6 that 35% of all the papers published within a year have no citation followed by 49% of all papers published having only a single citation. With just these two categories we see that 84% of all scientists have papers that are cited at most once. Those papers with only 2 citations account for the next 9%; those with just 3 citations included the next 3%; 2% have 4 citations; 1% with 5 citations and 1% with 6 or more citations. This distribution implies that 95% of all papers published are below the average of 3.2 citations/year in the number of times they are cited. Consequently, if a scientist's publications are cited 3.2 times on average, that person is a truly exceptional individual. His/her work is cited more frequently than 95% of their peers, so that rather than being a criticism, as used in the tenure committee argument, this is an indication of the superiority of the scientist's work.

Barabási [8] argues that inverse power-law networks like the World Wide Web (WWW) and the distribution of citations to scientific papers are scale-free[4] because the networks continuously expand by adding new documents to the WWW and the ongoing publication of new papers. Moreover each new node is preferentially attached to those nodes that already have a large number of links. New documents are linked to YAHOO, CNN and so on, while those papers that have been cited are more likely to cited again, snowballing in time.

Unlike Gauss' world of the bell-shaped curve, in Pareto's world of the inverse power law, the average achievement is truly exceptional. The average person surpasses the vast majority of those with whom s/he competes and

---

[4]While it is true that a class of inverse power law networks are scale free, it appears that the Internet is not one of them. Moreover there are multiple network architectures that yield inverse power-law structure, all of which scale, but are not all scale free.

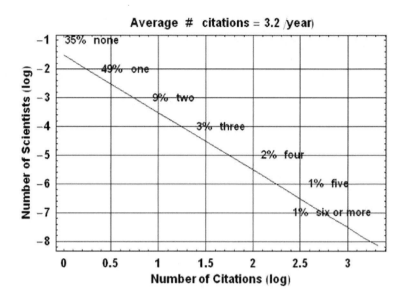

Figure 3.6: The distribution of the number of citations to all papers in the physical and mathematical sciences in a given year. The solid curve has a slope of -3 indicating that the distribution of citation number N decreases as an inverse power law of the form $(1/N^3)$. [modified from de Solla Price [161].]

that individual ought to be compensated/recognized in that proportion, or they would be if the world were fair.

The inverse power laws that quantify the properties of interest to the scientific community suggest that any attempt by management and administrators to equalize the rewards within a research organization will be counterproductive and disrupt the interactive modes that would ordinarily, adaptively develop between members. This *unnatural* equalization process is a consequence of the failure to recognize that the work of the scientific community is judged not only by managers, but more importantly, by a professional community of scientists.

Consequently, with the 80/20 rule in mind, complex networks suggest that managers, evaluating the overall quality of research organizations, should concentrate on the 20% that truly make a difference and only lightly monitor the 80% that do not significantly influence the direction of the research activity. Using this principle for guidance, the human network management of the scientific work force should take cognizance of what motivates scientists. For example, they are stimulated by interesting and challenging work, but only the top 20% are sufficiently self-motivated and talented to do the requisite basic research. The other 80% can often be guided to transition such fundamental research towards societal or military needs. The two groups are complementary to one another and a successful research organization requires both in order to generate innovative ideas, to carry out the research to test these ideas; to publish the results of the research in the appropriate journals to give talks at national and international meetings and file for the necessary patents. The 80% may be replaceable but they certainly cannot be done away with.

Again taking cognizance of the 80/20 rule it is clear that the 80% also desire high-caliber colleagues, for example, those who are well-published and whose work is well-cited in the scientific literature. The inverse power-law nature of the number of publications and citations indicates the level of impact the cited research has on the scientific community and draws attention from the greater community to the research activities of the laboratory. Such 20% colleagues provide a stimulating environment in which to do challenging research and the enhanced visibility of the organization increases the chances of the 80% to collaborate productively with scientists in industry and academia.

The successful management of the scientific network can only be accomplished by enabling the 20% to focus on 'unfettered' basic research who are unreservedly supported by the 80%. Management must bear in mind that this 20% does not always consist of the same scientists. From one year to the next any given scientist can be one of the 20% or one of the 80%, but only if the research environment is structured in such a way as to enable such mobility. It is management's willingness to allow for judgment-free change that ultimately determines the quality of the laboratory. Both scientific leadership and scientific support are important in an effective research laboratory and enabling one at the detriment of the other, in the end, serves neither.

The law of Pareto concerns the distribution of income and that of Lotka the distribution of scientific publications, but there are other laws with the same inverse power-law form: the law of Auerbach on the distribution of the sizes of cities, the law of Zipf on the relative frequency of words in language, the law of Richardson of the distribution of the sizes of war, the law of

Willis on the distribution in the number of gender in a species. All these distributions stem from the implicit multiplicative nature of the underlying networks that make them amenable to the Pareto Principle. The essence of this principle is that a minority of input produces a majority of the results and this is the key to understanding the properties of collections of complex entities, such as scientists. The Pareto Principle requires that *the few be important*, and in the present context, these few determine the properties of complex research teams and the many are irrelevant, which is to say, readily replaceable with other scientists of comparable ability. The trick is to know which is which, and how research teams can be assembled to contribute to the science of networks.

3.2.3 Information networks

Digital technologies have resulted in an explosion of sensor hardware capability, allowing better data resolution and greater speed, which is certain to continue into the future. This recognition of large-scale data collection has catalyzed collateral research into advanced techniques of information extraction, which is to say, the detection of patterns within the data that constitute the information of interest. New statistical data analysis techniques place the computer in the role of *data analyst assistant*, so that the identification and extraction of information patterns is automated. The human enters the process to transform the information into useable knowledge. For example, the blips on the screen are synthesized into the pattern of a man-made object, which the combatant interprets as a tank and the air traffic controller interprets as an airplane.

Information networks are designed to transmit information with minimum distortion and maximum traffic, or said differently, maximum fidelity and minimum transmission time. This traffic control problem was solved by properly designing for the variation in the time interval between successive messages on the network. The first modern communication network was that for the stationary telephone, where a telephone was connected by a wire which went into the wall of a residence and out onto a telephone pole whose wires physically connected distant points by means of intermediary switching stations. The design focus of this network was on the time intervals between successive telephone calls and the statistics were found to be simple. The probability that the number of calls in a given time was a specific number was determined by the distribution of Poisson, which is one of those bell-shaped curves. In fact, in a certain limiting situation the distribution of Poisson reduces to that of Gauss. The network property that leads to this statistical description is that the likelihood of receiving a message in any given time interval is proportional to that time interval. In this way the average number of messages received at a switching station per unit time is a constant.

The architecture of the world's first telephone network was based on this simple statistical principle and it worked remarkable well. When the message traffic on the phone network became too heavy, calls had to be delayed, forming queues of various lengths at the switching stations. One could, with high accuracy, determine how long a customer had to wait before their connection was made on average. Knowing the queueing statistics it was possible to design the network to make the waiting time the shortest possible.

Of course, the telephone network is a physical infrastructure connecting human to human; the statistics of message traffic becomes very different when the network is machine to machine, as it is on the Internet. The statistics that describe how message traffic is transmitted on telephone networks is not applicable to the traffic on the more complex Internet. The traffic on the computer network can not be described by an average any more than the economics within a society can be represented by an average income. In fact the increased complexity is manifest in the transition from a bell-shaped statistical distribution to one that is inverse power law, just as we described earlier.

The fusion or joining of information for making decisions is determined by an architecture consisting of many levels. The fusion begins with the data from a variety of sensors, which could be video, acoustic, tactile, electromagnetic as well as others. These different signals identify and characterize the state of single objects and includes the tracking of these objects over time. These various data sets are fused into information, which describe groups, objects or scenes, based on inputs from multiple sensor networked databases. This level is often called information integration, scene understanding, or situational awareness. The information from this level is used by the military in threat and impact assessment, to achieve the goals of faster, more effective missions at lower cost and with fewer casualties, using dynamical, interactive, stochastic algorithms. Finally there is sensor tasking, which through an understanding of objects, their relationships, their aggregate threat and possible impact, determine how sensors may be rearranged to enhance the information fusion process.

In addition to its role in transforming raw data into useful representations, information fusion facilitates the setting of prioritizes and how information is disseminated. Information must be human consumable, that is, information must be presented in a form that facilitates its transformation into knowledge.

On a more general note one of the main challenges for decision scientists in the twenty-first century is the managing of networks of ever increasing complexity. As networks such as electrical power grids, computer networks, and the software that controls them all grow increasingly complex, fragility, bugs, and security flaws become increasingly prevalent and problematic. It is natural to be concerned over the consequences this growing complexity has on our ability to manage these networks. Much of this complexity has to do with the things we create ourselves and the complexity is associated with the fact that we cannot represent the data in simpler or more compact form [63].

A challenge for network science, which will be met over the next two decades, is the development of techniques by which information is transmitted in a secure fashion to the appropriate level of information fusion in decision making. One such technique determines how one complex network transfers information to a second complex network and which has been applied to the phenomenon of global warming with interesting results [185], as we subsequently discuss.

3.2.4   Social-communication networks

The Harvard social psychologist Stanley Milgram (1933-1984) performed a social experiment in the middle 60s in which he mailed 160 letters to strangers in Wichita, Kansas and Omaha, Nebraska. In this letter he requested that these strangers contact by mail a certain stockbroker in Boston, but only by mailing the letter forward to someone they knew personally and instructing that person to contact the stockbroker through someone they knew personally. This was all explained in the letter, including the fact that they were participating in a sociological experiment. Milgram found that, on average, there was a chain of six people separating the strangers in Nebraska from the stockbroker in Boston. What this means is that of the people responding there was an average chain of six people connecting the person that received the letter to the stock broker in Boston. A number of variations of this experiment have since been carried out by Milgram and others and the results have been encapsulated into the phrase *six degrees of separation*, indicating that any two individuals are separated by no more than six other individuals. This is actually a metric for the 'distance' between people in a complex social network.

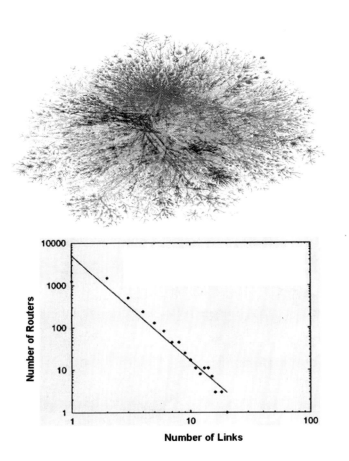

Figure 3.7: Top: a map of some 100 000 Internet routers and the physical connections between them. It reveals that a few highly connected nodes keep the network together. (Picture credit: W R Cheswick/Bell Labs). Bottom: the distribution of Internet nodes (routers) is depicted according to how many links each node possesses. The solid line is a simple inverse power-law fit to the data. [from [48] with permission]

The essential features of the social experiment have been modeled mathematically over the past few years and forms the core of what is now called Small World Theory in physics. This theory is reviewed, along with its history, in the books *Linked* by L. Barabási [8] and *Nexus* by M. Buchanan [19]. The version of small world theory presented in these popular accounts has become part of the intellectual mythology of network theory. Unfortunately the truth of this historical record seems to be in the eye of the beholder. The initial study made by Milgram, and often quoted in the retelling of the history, had fewer than 5% of people responded. With this rate of response the results were not statistically significant and probably for this reason Milgram never published the original study in the scientific literature. A second study, which was published, has a number of biasing factors and did not consist of the randomly selected group that is typically discussed [74]. The most dramatic differences between small world theory and the empirical historical evidence are the barriers presented by social classes to the linking problem. Consequently the mathematics of small world theory must be extended to account for the segregation of classes and social psychologists have yet to establish the empirical basis for small world theory.

Certain properties of small world social networks have been postulated to be generic and have been applied to modeling the Internet. The physical backbone of the Internet is given by routers and physical connections that are added to the network one by one. In this way the planet is developing a kind of nervous system consisting of computers, routers and satellites as nodes and phone lines, TV cables and electromagnet waves as links. The electronic nervous system forms a communications network with different kinds of components and a variety of connection types. It should be borne in mind that the Internet, as a whole, was not designed, but grew according to some principle that is only now revealing itself through the network's self-similar (scaling) properties, as depicted, for example, in Figure 3.7.

By juxtaposing the social and communication networks in this way we intend to emphasize how similar we believe the two be. This is most clearly evident in modern terrorist networks, given their reliance on the technology of the world they seek to destroy. Also in this dramatic example of a social network formed by terrorists the number of connections among the members forms an inverse power law. In Figure 3.8 the connections between the hijackers involved in the September 11, 2001 attacks on the Trade Towers in New York City are indicated. This network was reconstructed by Valdis Krebs using public information taken from major newspapers [76]. This ought to give the reader some indication of the information that is available to those that are suitably diligent.

The network immediately associated with the hijackers has $N = 34$ nodes, representing 19 hijackers and 15 other associates who were reported to have had direct or indirect interactions with the hijackers for a total of 93 links. In this map of the network the hijackers are color coded by the flight they were on and the associates of the hijackers are represented by dark gray nodes. The gray lines indicate the reported interactions with thicker lines indicating a stronger tie between two nodes. One can determine that for the terrorist network connected to those that committed suicide/murder in the attack there are 24 with $1 \leq k \leq 5$; 14 with $6 \leq k \leq 10$; 2 with $11 \leq k \leq 15$; 1 with $16 \leq k \leq 20$ and 1 with $21 \leq k \leq 25$. For a relatively small network the inverse power law is a good approximation to the distribution of connections.

The inverse power-law form of the network distribution has a number of implications for the stability and robustness against failures, whether through negligence, chance or attack. An understanding of these properties may provide guidance in protecting against failure in communication and information networks, as well as shut downs of the power grid. This understanding may also allow the formation of preemptive strikes at critical points in terrorist's networks. The distinction between robustness and fragility is crucial in this regard. The term robustness implies that the network continues to function even though the environment changes and its internal parameters also vary. The term fragile means that as a network parameter is changed, the parameter reaches a critical value at which a catastrophic change in the network occurs, often producing a failure without warning.

Many complex networks display an amazing ability to tolerate errors, and from the perspective of the network an error is a form of attack. This error tolerance was first explained in the context of physiology, where the stability of lung architecture was examined [183]. If the bronchial airways in the mammalian lung scaled classically, the size of an airway would decrease by a fixed percentage between successive generation of the bronchial tree. If this percentage was susceptible to fluctuations in the environment during morphogeneses, then the average diameter of the airway would increase with generation number as shown in Figure 3.9. By the fifteenth generation the error in the perturbed airways would produce an average diameter that is 3 to 4 times that of a healthy individual; the bronchial tree responds too strongly to errors in genetics, chemistry and the environment and therefore classical scaling would be maladaptive. On the other hand fractal scaling

Figure 3.8: The network surrounding the tragic events of September 11th 2001. The hijackers and their network neighborhood — directly and indirectly connected associates. Taken from Krebs [76].

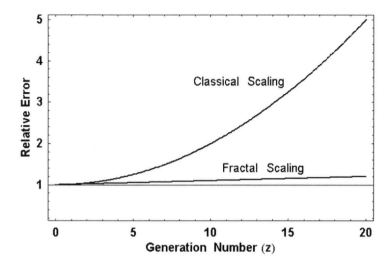

Figure 3.9: The error between the model prediction and the prediction with a noisy parameter is shown for the classical scaling model and the fractal model of the lung [from [183] with permission].

provided a relative insensitivity to errors of all kinds; at the twenty-second generation the relative error is less than 10%.

The fractal model of the lung is very tolerant of the variability in the physiological environment. This error tolerance can be traced back to the broadband nature of the distribution in scale sizes of a fractal network. This distribution ascribes many scales to each generation in the bronchial tee, consequently the scales introduced by the errors are already present in the fractal (inverse power-law) distribution. Thus, the geometrical structure of the lung is preadapted to a complex dynamical environment [187, 183].

The scaling nature of complex networks can make them less vulnerable to random attacks. The same arguments used for anatomical structure can be applied to the topology of the Internet, where a number of investigators have found that the structure is robust against random attacks. In this context the 80/20 principle implies that certain nodes in the network have many more connections than the majority of nodes, the hubs, therefore during a random attack it is one of the *majority* that is more likely destroyed and not one of the *few*. However, it is also clear that if the few can be identified

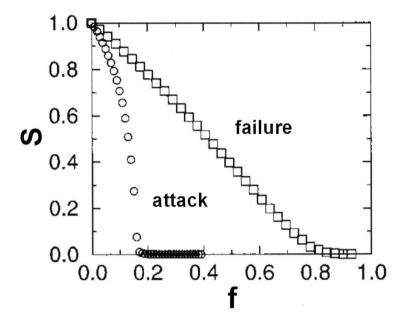

Figure 3.10: The deterioration in the size of the complex network having the scale-free topology. The causes of the reduction in size are random failure of nodes (squares) and directed attacks at hubs (triangles) [from [3] with permission].

and attacked selectively, then the network would be subject to catastrophic failure. This is the fragility of the scale-free network. The small world model of complex networks and its extensions have enabled scientist to test hypotheses concerning how inverse power-law networks respond to the failure of nodes using multiple strategies. In Figure 3.10 the response of a scale-free inverse power-law network to two kinds of disruptions is depicted.

The effect of random failure of nodes within an inverse power-law network is seen to be a linear degrading as the frequency of the failure is increased. However, when the network is being attacked and those attacks are directed at hubs a catastrophic collapse of the network occurs almost immediately, which is to say, for a very low frequency of attacks. This calculation done on mathematical models also implies how terrorist networks may respond to various kinds of attacks, or put the other way around, how scaling networks in western countries would respond to terrorist attacks. From the other perspective these calculations may also suggest how such networks may be designed to be robust against such attacks. In this way a scale-free network may become a scale-rich network and avoid catastrophic failure. Of course, other weaknesses may be found in the new networks, resulting in new designs.

3.2.5 Networks of neurons

Scientists have developed a number of mathematical models of the workings of the human brain. Some models emphasize that these workings require coordination among a scattered mosaic of functionally-specific brain regions that underlie an individual's ability to attend, perceive, think, decide, remember and act. Research shows that this coordination within the brain is a self-organized dynamical network that emerges spontaneously as a consequence of the nonlinear interaction among participating components.

The spontaneous activity of the brain is characterized by partially coordinated states in which component activities (individual neural regions) are neither completely independent of each other (segregated) nor linked in a fixed mutual relationship (integrated). Changes in the dynamical balance between the coupling among neural ensembles and the expression of each ensemble's intrinsic properties may be seen to create meaningful information. Thus, at least some cognitive processes appear to arise as the result of transitions to fully coordinated states that take the form of phase-locking synchrony in different frequency bands.

The modeling done by Kelso [71], for example, differs from the efforts of other scientists working in this area in that he believes that the biophysical coupling of neural populations, themselves composed of ensembles of Hodgkin-Huxley conductance-based neurons,[5] do not have attractors as solutions to the model systems of dynamical equations; or, at least, that the brain does not operate in the vicinity of such attractors. This lack of attractors in the brain's operating region implies the absence of states in the brain, so that in the brain's functioning only transients exist. Consequently there are only ghosts of the mathematical states through which the brain passes in its operation, stopping in no one configuration for any substantial

---

[5]This is a well-established mathematical model of the phenomenology of how electrical pulses propagate along neurons in the central nervous system.

length of time. This picture of brain dynamics emphasizes flexibility and adaptability and indicates that the same behavior can be produced by quite different histories.

A psychologist would say that an individual displays tendencies, whereas the mathematics would represent such tendencies by the coexistence of convergence to and divergence from non-stationary transient processes. This non-stationarity is a consequence of there being no stationary states for the brain, with the convergence and divergence corresponding to integrating and segregating the information.

In studying the dynamics of the brain we are at the peak of the complexity curve in Figure 1.10, where we do not have any metrics for the non-stationary nature of brain activity. We must look to the mathematics of networks, for describing the dynamics of the brain, in which the notion of a stationary state is replaced by a more functionally useful idea of a ghost state. The self-organizing behavior of coordination describes the tendency towards and away from these ghost states. This is truly a process to which the concept of equilibrium is alien, and although the brain may rest, it never stops.

3.3   Entropy and data processing

Many phenomena/networks are singular with no simple analytic representation because they do not posses a fundamental scale and require a new way of thinking to understand them. To use a military example, the terrain on which a battle is fought typically consists of sand, stones, rocks, boulders, hills, and mountains, with or without the foliage of grass, reeds, leaves and trees. The mix of these elements is such that no one scale characterizes the environment and traditional attempts to model the physical battle space require multi-scale dynamical elements. On the other hand, fractals provide the geometry for such natural settings and suggest new tools for finding manufactured objects in such settings. Such an object would be the only regular structure in the environment.

Another example of the new way of thinking can be demonstrated through an examination of how pervasive the notion of linearity is in our understanding of the world. Consider the simplest of control systems, that of aiming a weapon. When the aim of the weapon does not intersect the desired target's trajectory at the appropriate time, there is negative feedback to the input in direct proportion to the amount of error, to adjust the aim. Any hunter knows to lead the flying target so that the projectile reaches the point where the goose will be at some fraction of a second in the future. The mathematics of such simple tracking systems was developed in the 1940s and have been extended and refined extensively over the last half century. What has not changed is the way error and the adjustments made to reduce that error to zero are modeled mathematically.

In large-scale networks things are different. Many subnetworks have access to different information and must make their own local decisions, while still working together to achieve a common network-wide goal. Each subnetwork operates with limited information about the structure of the remainder of the network, which may dynamically reconfigure itself due to changing requirements, changes in the environment or by failure in other components. Controllers are required to be robust with respect to modeling uncertainty and to adapt to slow changes in the network dynamics. These considerations make the modern complex networks more like a living network than the mechanical tracking device of a gun.

Casti [23] argues that there is a one-to-one match between the problems that control engineers see as significant in complex networks and that biologists see in the problem of life, and this match up demands a reformulation of the standard modeling paradigm of control theory. Casti points out that what is required is extending the conventional input/output and state-variable frameworks upon which modern control theory rests to explicitly take into account the processes of self-repair and replication that characterizes all known living things.

Power laws are the indicators of self-organization, so that inverse power laws become ubiquitous, with order emerging from disorder in complex networks. In the transition from disorder to order, networks give up their uncorrelated random behavior, characterized by average values and become scaling, where the network is dominated by critical exponents. It is the exponent that determines the rate of fall-off of the inverse power law and captures the global character of the network, the imbalance repeatedly noted in the phenomena discussed earlier in this chapter.

It is evident from Figure 2.1 that in a fair world all the action takes place in the vicinity of the peak of the bell curve (the average value), since a random network has no preferences. In the inverse power-law network the action is spread over essentially all the available values to which the process has access, but in direct proportion to the number of links already present at each interval of time. This preference of a new member to hookup with an old member having established connections is what makes a star. A luminary in science is determined by social interactions in substantially the same way as is a luminary in Hollywood. The more a person is sought out, the greater the stardom.

Here we have introduced a number of definitions of entropy and argued that information entropy can be useful in characterizing the complexity of networks. The utility of information in measuring complexity is due, in part, to the fact that it is a physical, rather than a metaphysical, property of a network. The physical nature of information enables one to quantify the useable information associated with a complex phenomenon, in the same way that entropy quantifies the useable energy associated with the network.

In networks with only a few variables interacting nonlinearly, the dynamics are chaotic, generating free information. The amount of free information increases as the number of variables increases. This is represented schematically by the curve on the left-hand side of Figure 1.10, only now the measure of complexity is free information. For a very large number of variables, represented by the curve on the extreme right-hand side of Figure 1.10, there is no free information; there is only bound information at thermodynamic equilibrium. As the variables are identified as belonging to the network of interest or the environment, free information becomes available. With evermore free information being available with more variables in the network of interest interacting with the environment. At the maximum of complexity the network of interest has the maximum amount of free information. Moving to the right from the peak transforms free information to bound information and moving to the left from the peak less free information is being generated.

3.3.1   Diffusion of Information

Diffusion is one of those concepts developed in the physical sciences that became so completely understood that scientists from other disciplines were comfortable with making analogies to model very different phenomena. It is probably not surprising that our understanding of diffusion began with one of Einstein's 1905 papers, as we mentioned earlier. The phenomenon of physical diffusion was indeed first observed by the famous ancient Roman poet Titus Lucretius Carus (99 BC-55 BC). In his famous scientific poem *On the Nature of Things*, he noticed

> Observe what happens when sunbeams are admitted into a building and shed light on its shadowy places. You will see a multitude of tiny

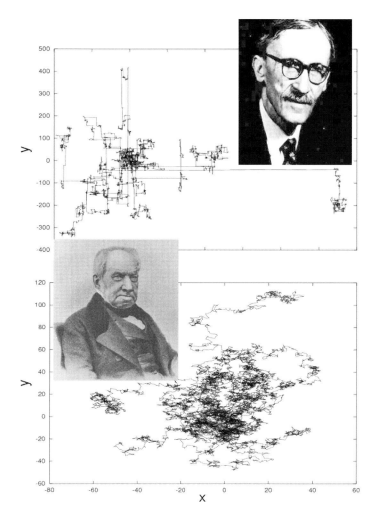

Figure 3.11: (Top) Paul Pierre Lévy (1886-1971) and an example of Lévy flight walk diffusion of 10 particles moving from the same origin with Pareto index $\mu = 2.5$. Note the typical island structure of clusters of smaller steps connected by a long steps or *flights*. The trajectories are statistically self-similar. (Bottom) Robert Brown (1773-1858) and an example of Brownian diffusion process, known also as *random walk*, of 10 particles. Long steps are very rare because the underlying statistics is Gaussian.

particles mingling in a multitude of ways... their dancing is an actual
indication of underlying movements of matter that are hidden from
our sight... It originates with the atoms which move of themselves
[i.e. spontaneously]. Then those small compound bodies that are
least removed from the impetus of the atoms are set in motion by
the impact of their invisible blows and in turn cannon against slightly
larger bodies. So the movement mounts up from the atoms and
gradually emerges to the level of our senses, so that those bodies
are in motion that we see in sunbeams, moved by blows that remain
invisible.

Although Lucretius mistakenly believed that the phenomenon he described was due to the motion of atoms, indeed in this specific case the
motion was due to air currents, his intuition surpassed his actual scientific
understanding of the phenomenon. The diffusion process was later rediscovered by the physician Jan Ingen-Housz (1730-1799) and subsequently by the
botanist Robert Brown (1773-1858), neither of whom could understand the
peculiar erratic motion of large suspended particles in a medium of smaller
particles. For Ingen-Housz it was charcoal suspended in alcohol, for Brown
it was pollen motes in water, but in both cases the trajectory of the large
particle was an erratic path whose speed and direction could not be determined from one point in time to the next. The first mathematical explanation came from Thorvald Nicolai Thiele (1838-1910), but it was Einstein
who introduced the phenomenon to the physicists and explained that the
motion of the heavy particle was a consequence of the imbalance in the
forces produced by the random collisions of the lighter particles, that is in
such examples the molecules of water, with the surface of the heavy particle. The instantaneous imbalance in the surface forces, pushes the mote
around like the suppressed motor control network of a drunken sailor and
subsequently this behavior became known as a random walk or a drunkard's
walk.

This is the phenomenon of classical diffusion in a homogeneous, isotropic
fluid. The lack of directionality in the fluid results in the average position of
the heavy particle not changing in time, even though the variance around this
average increases linearly with time. The probable position of the particle is
described by the bell curve of Gauss, centered at the initial position of the
particle and a width that increases linearly with time. If the particle has an
electric charge, then an external electric field can be applied to produce an

average drift, so the central region of the bell curve migrates in the direction of the external force.

In 1983 the Franco-American mathematician Benoit Mandelbrot (1924-) introduced the concept of fractal geometry [103] into science. Mandelbrot emphasized that fractals are realistic and useful models of many phenomena in the real world that can be viewed as highly irregular and which cannot be described by traditional Euclidean geometry. In fact,

> ... clouds are not spheres, mountains are not cones, coastlines are not circles, and bark is not smooth, nor does lightning travel in a straight line.[103]

Thus, typical examples of natural fractals include geophysical networks such as mountains, coastlines and river basins; biological networks such as the structure of plants, blood vessels and lungs; astronomical networks such as the clustering of galaxies; anomalous diffusions such as Brownian motion and Lévy flights, the latter being generalizations of classical diffusion to the situation where the variance is not finite: see Figure 3.11. There are also human-made fractals such as stock market prices, music, painting and architecture. In all these phenomena and many more it is possible to find scaling laws. Evidently, the theoretical and experimental search for the correct scaling exponents is intimately related to the discovery of deviations from ordinary statistical physics and of the important characteristics underlying a given phenomenon.

One of the most common scaling exponents for characterizing fractal time series is known as the *Hurst exponent* and is commonly indicated with the letter H. The scaling exponent H was coined by Mandelbrot [103] in honor of the civil engineer Harold Edwin Hurst (1880-1978) who first understood the importance of scaling laws to describe the long-range memory in time series [68, 69]. In particular, Hurst was interested in evaluating the persistence of the levels of the annual floods of the Nile River and developed a method of time series analysis that enabled him to express that persistence in terms of the scaling parameter H: see Figure 3.12. The memory in these time series is determined by the autocorrelation function that quantifies how much a given fluctuation in the time series influences another fluctuation that occurs a time t later.

Mandelbrot extended the work of Hurst and developed the concept of *fractal Brownian motion* or, equivalently, the concept of *fractal Gaussian*

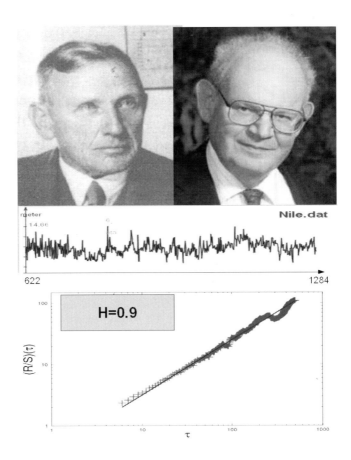

Figure 3.12: (left) Harold Edwin Hurst (1880-1978) and (right) Benoit Mandelbrot (1924-). The original sequence of the Nile River minima series sampled in yearly interval starting from 622 AD to 1284 AD used by Hurst for its scaling analysis that yielded a scaling function of the type $f(\tau) \propto \tau^H$ with a Hurst exponent of $H = 0.9$. Because the scaling exponent is between 0.5 and 1, the Nile River is proven to be a complex network regulated by long-range correlations.

noise for fractal time sequences.[6] These sequences differ from the short-range correlated time series that are characterized by an autocorrelation function that decays exponentially and which appears random to the Hurst analysis.

A value $0 < H < 0.5$ corresponds to anti-persistent[7] noise; $H = 0.5$ corresponds to uncorrelated or random noise, also known as white noise; and $0.5 < H < 1$ corresponds to correlated or persistent[8] noise. A value $H = 1$ corresponds to 1/f-noise or *pink* noise. The adoption of a color name derives from the fact that a light source characterized by a *1/f* spectrum looks pink. This type of noise is particularly important because it represents a kind of balance between randomness and order, or between unpredictability and predictability. In fact, for pink noise the autocorrelation between two events separated by a given time interval is constant, that is, the autocorrelation function is independent of the time interval.

The idea of a process being determined by the random influence of the environment was appealing to a number of social scientists. One application of the diffusive transfer of information is to a model of the propagation of rumors leading to the movement of prices in the stock market. Rumor mongers can be amateur analysts or individuals with access to inside information. These individuals usually do not have sufficient funds to influence the price of stocks through buying and selling, but given a sufficiently large audience, they may influence a stock's price. As pointed out by von Bommel, in recent years the Internet can provide a significant incubator for rumors and the most significant rumors are those that actually contain some information. Of course, there are also the fraudulent rumors based on misinformation or direct lies. But the former is significantly more prevalent than the latter. Of course rumors migrate in other social contexts, not just among people interested in the dynamics of the stock market.

In a social network information is passed from element to element (person to person) by direct interaction. This nearest neighbor hand-off is equivalent

---

[6]Fractal sequences are those characterized by an autocorrelation function $C(t)$ that decays as an inverse power law in time $C(t) \propto 1/t^\alpha$ with scaling exponent $\alpha = 2 - 2H$ or, equivalently, with a power spectral function $S(f)$ that decays as an inverse power law in frequency $S(f) \propto 1/f^\beta$ with scaling exponent $\beta = 2H - 1$. Here H is the Hurst exponent of the sequence.

[7]Anti-persistent noise means that if a value of the sequence is *up* the following value is likely to be *down*.

[8]Persistent noise means that if a value of the sequence is *up* the following value is likely to be *up* again.

to the force mechanism producing diffusion in the classical analysis. The random imbalance of impact forces on the heavy particle, producing its erratic motion, is analogous to the random movement of a packet of information through a network of people. Of course, this packet is left ill-defined as is the network through which it moves; the packet could be an infection for the spread of disease; or a new idea for the transfer of technology or innovation between companies; or perhaps a fad/fashion which can change the social environment.

An information packet can move randomly in any direction producing an increased uncertainty in location or in the time of arrival, and measures of that uncertainty grows in time. If there is memory associated with the movement the variance (uncertainty) no longer grows linearly, but algebraically in time, with power-law index H. The information spreads more rapidly than classical diffusion if $H > 1/2$ and less rapidly if $H < 1/2$. Consequently, H provides a measure of how erratic the propagation of information is within the network. Suppose the information is the time from one heart beat to the next, as in the HRV data discussed in Chapter 2, so that H is related to the scaling index displayed in Figure 2.8. This measure of the cardiovascular network provides a way to characterize the variability of heart rate. In normal healthy adults the typical range of the Hurst exponent is $0.7 < H < 0.9$, indicating that information diffusion is persistent [187]. As people age the index approaches the classical diffusion value of one-half, which is to say, the cardiovascular memory diminishes as we age.

3.3.2   Diffusion entropy analysis

We have now established two things. First, entropy has been used as a measure of information for over half a century and so the underlying theory is well developed, both in terms of the underlying mechanisms and the range of applications of that theory. Second, the connectivity for complex networks measured in the real world are inverse power-law or heavy-tailed distributions, at least asymptotically. Putting these two findings together we seek to determine how the entropy measure of information can be exploited for the understanding of complex networks described by inverse power laws.

Scafetta and Grigolini [138, 139] recognized that fractal time series, which scale, are characterized by probability densities that scale. This scaling property immediately leads to a simple algorithm for estimating the Hurst exponent H as the scaling exponent of the standard deviation (or variance). This is accomplished by constructing a diffusion process from the elements of the time series by aggregation. This procedure of aggregation, or adding discrete measurements together, transforms the description from that of a discrete data set into that of an erratic trajectory, or more exactly into a collection of erratic trajectories. This method has been called Standard Deviation Analysis (SDA). Several alternative methods for evaluating H based on second moment analysis have been suggested: various comparisons of these methods have also been made [139].

However, the fact that fractal time series are characterized by probability densities that scale makes possible a methodology that provides an alternative to those based of the second moment. The probability density function of the diffusion processes generated by the above sequences can be evaluated and provides a second scaling exponent δ.[9] An efficient way to get at this new exponent is through the information entropy that is constructed using the probability density. The method is called the Diffusion Entropy Analysis (DEA). As in the SDA, in DEA the time series is aggregated to transform the description from a discrete data set into a collection of trajectories. The probability density that characterizes this ensemble of trajectories is then used to calculate the entropy for the time series. In Figure 3.13 is sketched the scaling behavior of three kinds of ensembles. The lowest one, having a slope of $\delta = 0.5$, is that of a completely uncorrelated random process corresponding to a particle diffusing in a fluid of lighter particle like cream in a morning cup of coffee. In this case the new scaling exponent is the same as that determined by the second moment. The uppermost curve, having a slope of $\delta = 1.0$, is that of a completely regular process undergoing ballistic motion, like a bullet fired from a rifle. Between these two extremes of motion fall the cases of interest, those being complex networks in which the slope of the entropy determines how close the network of interest is to the two

---

[9]If the time series is sampled from the random variable $X(t)$ and the data scale such that the probability density satisfies the scaling relation

$$p(x, t) = \frac{1}{t^\delta} F\left(\frac{x}{t^\delta}\right)$$

which when substituted into the definition of the entropy yields

$$S(t) = S(1) + \delta \ln t.$$

Consequently, graphing the entropy versus the logarithm of the time yields a straight line whose slope is the scaling parameter δ.

extremes where $\delta \neq H$. One type of network that falls in this middle domain is one having long-time correlations.

The information contained in the time series of interest can be extracted by calculating the entropy using the probability density for the underlying process by means of the DEA and the scaling of the standard deviation of the same diffusion process by means of the SDA. However, why do we need two methodologies? What is the relation between the scaling exponents H and δ?

To answer the above question the reader should know that there exist at least two different kinds of complex fractal sequences: Fractal Gaussian noise and Lévy flight noise. These noises are characterized by different statistics according to whether the diffusion process generated by them obey the *Central Limit Theorem* or the *Generalized Central Limit Theorem*. In the former case, the diffusion process generated by the sequence converges to a Gaussian probability function, in the latter case the diffusion process converges to a Lévy probability function.[10] Lévy processes were introduced by a French mathematician, Paul Pierre Lévy (1886-1971), who was actively engaged in exploring the weaknesses of the central limit theorem.

In the case in which the diffusion process converges to a Gaussian function, because in this case the probability function scales like the second moment, we have that the two scaling exponents are equal: $H = \delta$. In the case in which the diffusion process converges to a Lévy probability function, the second moment diverges, so that the scaling exponent H is not defined[11] while the DEA exponent δ is related to the Lévy exponent μ as $\delta = 1/(\mu - 1)$.

Fractal Gaussian noise and Lévy flight noise are very different. The former derives its anomalous scaling from long-range correlations and the corresponding probability density may be any analytic function with a finite second moment. One way to determine if this is the nature of a given data set is by shuffling the data, that is, randomly changing the order of the data points in time. Since no data is added or subtracted in this process, the statistics remain unchanged, however the scaling of the second moment now has $H = 1/2$ corresponding to an uncorrelated random process. Any correlation in the original data is destroyed by the process of shuffling. These data

---

[10]A Lévy probability function is characterized by tails that decrease as an inverse power law with exponent $2 < \mu < 3$.

[11]Numerically it is found that $H = 0.5$ if the Lévy sequence is random.

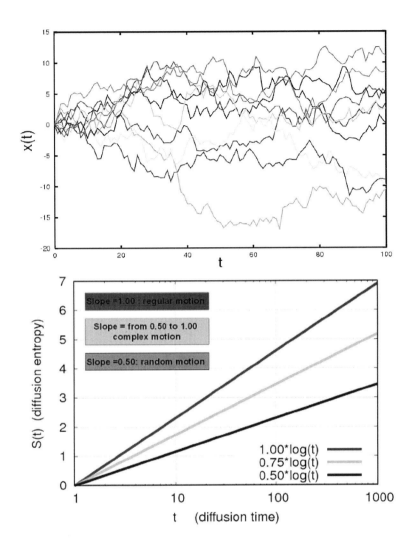

Figure 3.13: (Top) Diffusion trajectories of a random walk process against the time. (Bottom) The three curves represent networks whose probability densities scale in time. The lowest is that of uncorrelated random motion; the uppermost is that of completely regular motion and the middle curve, labeled complex network, blend properties of the two in the networks dynamics.

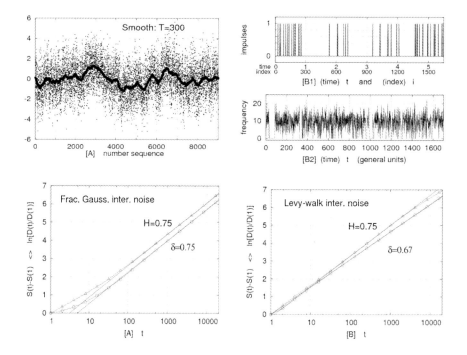

Figure 3.14: (A/top) Fractional Gaussian noise with H=0.75; (B/top) two forms of Lévy walk intermittent noise with waiting-time distribution $\psi(\tau) = 1/\tau^{\mu}$ and $\mu = 2.5$ (B2 gives the frequency of impulses every 300 units of B1). (Bottom) DEA and SDA of the above signals, respectively. Note that for fractional Gaussian noise the scaling lines are parallel, that is, $\delta = H = 0.75$. Whereas for the Lévy walk the two scaling lines diverge with $H = 0.75$ and $\delta = 0.67$ in accordance with the Lévy-walk relation for $\mu = 2.5$.

are called fractal Gaussian noise because the central limit theorem would yield the statistics of Gauss for a sufficiently long time series and fractal because of the anomalous scaling of the second moment $H \neq 1/2$.

Lévy flight noise, instead, derives its anomalous scaling from the shape of its probability distribution function that has inverse power law tails. Like fractal Gaussian noise, shuffling the data described by Lévy flight noise does not change the properties of its probability distribution function, however, unlike H for the second moment, the scaling parameter δ would remain unchanged by such an operation. For these reasons it is usually straight forward to distinguish fractal Gaussian noise from Lévy flight noise, just look at the probability distribution function of the data.

However, there exists a particular kind of noise that looks like Fractal Gaussian noise because its probability distribution function has a finite variance, but its statistics obeys those of Lévy and not of Gauss. These special noises are called *Lévy walks*, that are characterized by having an underlying waiting-time distribution of the form of a Lévy function, such as the one we have found for the occurrence of the solar flares (see Figure 2.15). For these special noises both H and δ can be calculated, but they differ from each other, and are related by the Lévy walk scaling relation: $\delta = 1/(3-2H)$ [139]. Because, Lévy walk signals are often disrupted by noise that easily suppresses the Lévy waiting-time tails, their Lévy statistics remains concealed from the traditional variance-based analysis used by most investigators. Such data sets are easily mistaken for fractional Gaussian noise unless revealed by the combined measurement of two complementary scaling exponents H and δ using SDA for the former and DEA for the latter.

Figure 3.14 shows a comparison of fractal Gaussian noise and Lévy walk noise artificially generated with a computer. The Lévy walk noise is produced in two forms: as a continuous intermittent signal and as a discrete signal obtained by averaging the first one. Note the data in Figure 3.14B2 where the data look just like fractal Gaussian noise but, indeed, they are by construction a Lévy walk noise. The entropy and the standard deviation are calculated by means of DEA and SDA using the time series, and the results are graphed in the figure as a function of the length of the diffusion time. It is evident from the figure that the two scaling exponents are equal in the case of Fractal Gaussian noise and are different in the case of Lévy walk noise. In the latter case the two scaling exponents obey the Lévy walk scaling bifurcation relation: $\delta = 1/(3 - 2H)$. The figure stresses the importance of combining the traditional Hurst analysis with the entropic approach for distinguish two signals generated by very different anomalous fractal statistical mechanisms that correspond to different dynamics.

3.4   Sun-climate complexity matching?

When one complex network perturbs a second complex network, it may be possible to deduce certain general properties of the response based on the relative values of the measures of complexity. Thus, Scafetta and West [140, 143] hypothesized that the Earth's average surface temperature may inherit its anomalous fractal statistics from the Sun. The hypothesis was based on the fact that the Lévy nature [61] of solar flare intermittency, which is characterized by two scaling exponents (δ and H as defined in the previous section), can be extracted from the global temperature data sets as well. The solar flare intermittency can be assumed to mimic the dynamics of the total solar dynamics. Thus, the scaling exponents characterizing solar flare intermittency, as derived from the diffusion scaling algorithms, are representative of the total solar dynamics as well. The mathematical details relating these scaling parameters are discussed in the technical journals.

In Figure 3.15 SDA and DEA are applied to the temperature data and to several solar records. The figure shows that the scaling coefficients H and δ of several solar data and global temperature time series and the exponent $\mu = 2.12 \pm 0.05$ of the waiting time of solar flares are consistent with the Lévy walk relation. Thus, Figure 3.15 suggests that the global climate inherits its peculiar fractal persistence and long-range correlation from the turbulent Lévy activity of the Sun.

We observe that the diffusion algorithm we adopted to estimate the scaling exponents can detect the long-range fractal persistence of a time series even if the actual data is a superposition of a long-range correlated signal and a more random component, as may be plausible in climate data. This separation of effects is possible because the diffusion process generated by the data is driven by the superdiffusion generated by the component of the signal with the strongest fractal long-range correlation. Therefore, the diffusion algorithm would emphasize the strong fractal solar-induced component of the global temperature and de-emphasize the more random climate components such as volcanic activity, and the smooth upward trend caused by a gradually increasing greenhouse gas concentration.

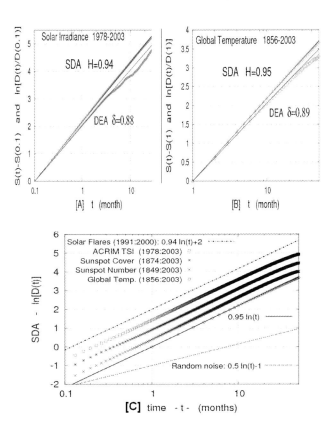

Figure 3.15: [A] DEA and SDA applied to the ACRIM composite TSI time series. The two straight lines correspond to the scaling coefficients $\delta = 0.88 \pm 0.02$ and $H = 0.94 \pm 0.02$. [B] DEA and SDA applied to the global temperature anomalies (1856-2003) time series. The two straight lines correspond to scaling coefficients $\delta = 0.89 \pm 0.02$ and $H = 0.95 \pm 0.02$. [C] SDA is extended to the sunspot number and cover records and we find $H = 0.94 \pm 0.02$ for both terrestrial and solar data. According to the Lévy walk relation, the inverse power exponent $\mu = 2.12 \pm 0.05$ of the waiting time of solar flares (see Figure 2.15) would induce a Lévy-walk with theoretical exponents $H_T = 0.94 \pm 0.04$ and $\delta_T = 0.89 \pm 0.04$. [Taken from [144]].

Therefore, in addition to the familiar 11-year and 22-year quasi-periodic solar cycles, the total solar irradiance (TSI) time series seems to present a fractal persistence of the statistical fluctuations that is consistent with the Lévy-like persistence that describes solar flare intermittency. The significance of Figure 3.15 is that the curves support the hypothesis made by Scafetta and West [140, 143], where only solar flare data were used, that an underlying Lévy-like temporal pulsing process regulates the evolution of total solar activity. The solar dynamics is then inherited by the Earth's climate as established by the equivalence of the scaling exponents between solar and terrestrial data in Figure 3.15.

If the above findings are not accidental (a conclusion we cannot rule out, but which does seem rather remote), they would imply the existence of a complex Sun-climate nonlinear linking, that could not be efficiently detected by linear methods of analysis such as multilinear regression analysis, nor by linear climate models. Therefore, the study of the Sun-climate linking may require the use of all potential mechanisms connecting solar variations to changes in the Earth's climate. Until this is possible, an alternative approach has to be investigated; an approach that exhausts all phenomenological connections between solar and climate dynamics.

In this and in the previous chapter we have seen that both climate and solar proxy reconstruction data are uncertain, but that the information we can infer from them is that there exist a linking between solar dynamics and global climate. We have seen this by means of simple sequence comparison such as in Figure 2.18 and historical records that suggest a Medieval Warming Period and a Little Ice Age that corresponds to what we can infer about secular solar activity. In this chapter we have presented evidence that a complex stochastic linking exists between solar data and the earth's climate. This linking suggests that the solar influence on climate may be quite significant. However, in the first chapter we reviewed a conclusion reached by a number of scientists concerning some important climate models that suggest a weak solar influence on climate. Thus, what is the *knowledge* we do have about Sun-climate interaction? In the next chapter, we show that this *knowledge* is itself a complex network that depends on several factors, including sociological ones, that can easily disrupt it. We expose this dependence by analyzing both traditional and phenomenological climate models and how these have evolved over time.

Chapter 4

Knowledge

If data is the outcome of experiment and information is the order contained in the data, then what is knowledge? It seems to us that knowledge is that ephemeral quality that extends over and above both these concepts enabling us to interpret the order in the data, without reference to a particular data set. Knowledge is the interpretation of the order in the data in the context of a theory.

For example, as we lower the temperature of water it eventually undergoes a transition from the liquid phase to the solid phase. This information about the phase transition of the water is understood in the context of the microscopic theory of matter, that is, that all matter is made up of microscopic particles that interact with one another. The ordering that occurs in the transition from liquid to solid arises because the particle interactions change their character. The interaction changes from being short range and dominated by erratic thermal fluctuations to being long range and dominated by collective behavior. It is theory that allows us to see the temperature dependence of the interactions and consequently how that dependence results in the phase transition. It is theory that enables us to extend our understanding of the particular and apply it to the general. It is the interpretation step that constitutes knowledge in physics in particular, and science in general.

In the first part of this chapter we attempt to extend this notion of knowledge beyond any particular discipline and associate it with the network underlying the phenomenon of interest. If there is to be such an entity as Network Science then it must supersede the traditional scientific disciplines, while at the same time be reducible to a discipline in the appropriate limit. In order to achieve this goal requires a mathematical description of the complex network being discussed, so that the dependence on the appropriate metrics can be made systematic. Consequently the distribution of Pareto that was shown to be ubiquitous in the previous chapter constitutes the information that we propose to interpret using a science of networks. Said differently, the inverse power laws of the various phenomena considered suggests that there may be an overarching way to represent the structure of the underlying networks that constitute a theory on which a Network Science may be based.

Thus, we examine a number of network architectures and in so doing find various ways to understand how different mechanisms give rise to the observed inverse power-law behavior observed in biology, botany, chemistry, sociology, physiology, physics, psychology, and zoology. We review some of the basic properties of complex networks, starting with the simplest, in which the connections among the nodes are constructed to be random. Random networks are typically static and are too restrictive to mimic the full range of behaviors of most real-world phenomena. Random networks are characterized by average quantities just as in the case of Gauss statistics. Two real-world properties that random networks do not capture are growth and a criterion for establishing new connections in growing networks. One mechanism is stated as a rule by which newly formed nodes prefer to make connections to nodes having the greatest number of existing connections within a network. This rule, along with growth, is sufficient to develop an understanding of a class of scale-free networks. Such scale-free networks have Pareto distributions in the number of connections to a given node. There is a deep relationship between these scale-free networks and the fractal statistics discussed previously, see, for example, as found in the case of physiology [187].

Scale-free networks are not the end of the story, in fact, they are only the beginning. An engineering approach to understand inverse power-law networks is based on design and has the acronym HOT, originally short for Highly Optimized Tolerance. Subsequently the acronym was used to mean Heavily Organized Tradeoffs and Heuristically Optimized Tolerance, and now it is used for any of a number of changes of the three words with the appropriate letters. The starting point for this approach was the work of Carlson and Doyle [22] in 1999 in which complex networks of highly engineered technology are viewed as the natural outgrowth of tradeoffs between network-specific objectives and constraints. The result is a *robust, yet fragile* network behavior.

The issue finally addressed in this chapter is how to use the scaling behavior of phenomena to control complex networks. Homeostatic control is familiar from medicine having as its basis a negative feedback character which is both local and free of time delays. The relatively new allometric control, on the other hand, can account for long-term memory, correlations that are inverse power law in time, as well as, long-range interactions in complex phenomena as manifest by inverse power-law distributions in the network variable.

In the latter part of this chapter we continue our discussion of global warming. We attempt to reconstruct how our knowledge about the increase in the Earth's average temperature has emerged during the last decade. We show that the *knowledge* itself of a physical phenomenon is a kind of *living* network that grows and evolves in time. This living network is not just additive: a network where new nodes and new connections are just added to the old ones. Sometimes old nodes and/or old connections have to die because of the newly added ones, and when this occurs the entire network may undergo a major reorganization. Such a change leads to a completely different understanding of the physical phenomenon. We illustrate this using the global warming debate.

In science, the simplest major disruptive event that may lead to a completely new reorganization of the knowledge network about a given phenomenon is given by, for example, the discovery that the data on which a previous interpretation was based are wrong. In the previous chapters we have seen that today both solar and climate data on a secular and decennial scale are very uncertain and still debated. However, the implications of this fact are still not well understood in the scientific community that bases its understanding of global warming on models that were built to agree with the previously available data reconstruction.

We briefly reconstruct the historical record of this debate and show how a basic phenomenological understanding was first overthrown by a theoretical climate model understanding of the phenomenon that produced simulations fitting the available data, in particular the *Hockey Stick* graph of the secular Earth's surface temperature. Subsequently, the publication of new reconstructions of the secular surface temperature and a detailed comparison between the data and the computer model simulations indicate significant limitations of the theoretical climate model approach. Finally, we show how more advanced phenomenological models interpret the data and suggest a significant solar effect on climate change contrary to what is claimed in the

IPCC report. Indeed, the ongoing debate on global warming seems to lead to *knowledge* that may be significantly reshaped in a few years.

4.1   Mathematics enables knowledge

Mandelbrot showed that his fractal concept appears everywhere, from neuronal trees, whose structure is described by Rall's Law [130], to botanical trees described by Murray's Law [116], and which started five centuries ago with the observations of Leonardo da Vinci [134] that successive branchings of a tree are related to one another by a simple mathematical relationship; what is now called scaling. This description of the static interconnectedness of geometric scales must now be adapted to dynamic complex networks.

It is important that we understand the reasons why present day mathematics/ science has such a hard time describing truly complex networks. It is only through this understanding that we can see what innovative mathematics/science is required for a science of networks. Novel ways of thinking are necessary to understand networks that have no analytic representation, that require scaling and that blend regularity and randomness.

The transformation of data into information and the subsequent transformation of information into knowledge involve the multiple interactions of structured networks to perform particular tasks. In general, the fundamental elements of which the cognitive and response networks of the individual are composed are themselves complex rather than simple; the interactions among the subnetworks, as well as within subnetworks, are generically not linear and therefore do not lend themselves to traditional stimulus-response analysis of psychological networks. Finally, the properties of subnetworks observed in the laboratory almost never scale without modification to field conditions. This need to understand the nature of complex networks, self-adaptive behavior and complexity itself has in part led to the development of information networks.

Biological networks have a number of properties in common with the most complicated of physical networks. In general, living networks have a large number of dynamical variables that interact nonlinearly with one another over multiple scales in space and time, producing time series that are always irregular and often chaotic. Consequently, traditional data processing techniques are inappropriate for the analysis of such networks because they suppress information (through filtering) by attributing erratic behavior to noise rather than to nonlinear dynamics. What is required is a systematic study of metrics for complex networks; metrics that quantify the information that had previously been filtered from the *signal* and therefore was lost to the analysis. For example, the heartbeat is a traditional measure of health, but recently it has been found that the variability in heartbeat, rather than the heartbeat itself, is a more robust measure of health [187]. Similar measures for breathing, human gait and even traumatic brain injury (TBI) have also been developed using the concept of variability. These measures of variability or others very much like them will be the health metrics of the future.

In the past two decades a great deal of research has gone into the development of multidisciplinary approaches to describe phenomenon that are not strictly within the realm of a single discipline. This was, in fact, how new disciplines were born. However, mathematical methods that characterize complex phenomena without regard for specific mechanism involved in the process may be called transdisciplinary. The measurable quantities associated with such mathematics can be called transdisciplinary metrics. One example of such a metric is the information content of a DNA sequence and the mathematics consists of the biocomplexity algorithm used to understand the calculation being done. Others could be the information in computational neuroscience, and metrics associated with an individual's situational awareness and cognitive readiness, based on physiological variability.

New mathematical tools are required to support evolutionary and revolutionary development in the modeling and simulation (MS) of complex phenomena. The MS theme directly supports the enhanced modeling and simulation capabilities to design, develop, acquire, support and utilize new networks. The ongoing research supports the modeling of complex networks through the development of techniques and methodologies that facilitate the modeling of materials, biological organisms, as well as mechanical and electronic networks. The development of mathematical enablers supports the art and science of simulation as is necessary to permit the efficient and effective use of simulations in support of teaching, scientific inquiries and analysis, visualization, decision making, enhanced predictive capabilities, and human behavior representation.

The descriptors of complex phenomena, now and in the future, include multi-scale and multi-resolution analysis of equations of motion. In addition, the analysis of complexity supports the research to identify, understand and mathematically formalize the metrics in complex physical, biological and informational networks in order to enhance the detection, identification and response to threats of survivability, loss of lethality and mobility, as well as to inhibitions of adaptiveness.

We anticipate that devices, artificially produced networks, and information processing may have to be optimized over an entire network, rather than within a subnetwork as is done now. This holistic approach would impact the characterization of materials, including advanced composites and smart materials. The decision-making in large organizations that consist of multiple departments having different purposes, but all supporting the overall goal of the organization. Finally, there is the issue of networks containing large numbers of sensing devices, issues of network organization and topology, because scientific principles have yet to be created that properly capture the properties of distributed information with long-term memory, that is, the influence across components are nonlocal in space and time.

Teletraffic theory is the mathematics of the design, control and management of public switched telephone networks. Historically this theory incorporates the discipline of statistical inference, mathematical modeling, optimization, queuing and performance analysis. This was later extended to the global Internet, but those that built the Internet did so without an appropriate teletraffic theory. The existing theory did not work for the Internet because the way computers talk to one another is different from the way humans talk to one another. Rather than being described by a Poisson statistical distribution, as are the arrival times of voice calls, Internet traffic is described as a fractal random process. In addition the properties of the individual sites change over time, so the entire network is non-stationary. All this leads to a new class of problems. Rather than the traditional sparse data problem, we require inference methods for large collections of high-volume, large throughput, non-stationary data sets.

Scientists have begun to understand the nature of computer networks through the application of Pareto distributions. These successes argue for the future application of fractal geometry, fractal statistics and, equally important, fractal dynamics, to the understanding of complex networks. One path to understanding is through the application of wavelets to uncover the multi-scale properties of networks and the development of new wavelet-based techniques. The research into new kinds of mathematics, such as the fractional calculus to describe complex dynamics and biocomputing, as well as innovative numerical integration techniques, shall provide new strategies for solving such long-standing problems.

Finally, complex networks require an original kind of mathematics for their understanding, and this new mathematics will, in turn, require a unique way of thinking. It was a revelation to the physics community that determinism does not imply predictability, meaning that just because one has the equations of motion, this does not mean that one can predict the future behavior of a given phenomenon. This was the lesson of chaos. Nonlinear dynamical equations have solutions that are so sensitive to initial conditions that it does not make practical sense to talk of predicting the final state of the system from an initial state with even an infinitesimal amount of error. This insight has been used to explain our failure, using traditional mathematical models, to describe classes of inhomogeneous chemical reactions, turbulence, physiological phenomena such as the beating of the heart, the mechanical properties of common materials like clay and rubber, and so on through a host of other natural and man-made complex networks.

4.2   The Pareto Principle: the 80/20 rule

The modern theory of human network management can be traced back to Joseph M. Juran who wrote the standard reference work on quality control [70]. It is this work on management that heralded the emphasis on organizational metrics and the importance of being able to quantify the problems being addressed within an organization. The Gauss distribution is often used to quantify the variability observed in manufacturing output such as achieving the machine tolerance on a *widget*. The applicability of such a measure to organizational output, such as the number of publications per year of a research laboratory, however, is less clear. The Gaussian view of the world treats networks as if they were made up of linear additive processes which Juran thought inappropriate for the management of people, that is, to social networks.

In itself, constructing a distribution of the inverse power-law type would not necessarily be particularly important. Parts of what makes the Pareto distribution so significant are the sociological implications that Pareto and subsequent generations of scientists were able to draw from it. For example, Juran identified a phenomenon that later came to be called the Pareto Principle, that being that 20% of the people owned 80% of the wealth in western countries, what was called the fundamental imbalance in our earlier discussion. It actually turns out that fewer than 20% of the population own more than 80% of the wealth, and this imbalance between the two groups is determined by the size of the Pareto index. The actual numerical value of the partitioning is not important for the present discussion; what is important is that the imbalance exists in the data.

In any event, the 80/20 rule has been determined to have application in all manner of social phenomena in which the few (20%) are vital and the many (80%) are replaceable. The phrase "vital few and trivial many" was coined by Juran in the late 1940s and he is the person that minted the name Pareto Principle and attributed the mechanism to Pareto. The 80/20 rule caught the attention of project managers and other corporate administrators who now recognize that 20% of the people involved in a project produce 80% of all the results; that 80% of all the interruptions in a meeting come from the same 20% of the people; resolving 20% of the issues can solve 80% of the problems; that 20% of one's results require 80% of one's effort; and on and on and on. Much of this is recorded in Richard Koch's book *The 80/20 Principle* [75].

These properties of complex networks are a consequence of the imbalance in society first quantified by Pareto. In corporations the imbalance in the distribution of achievement translates into the majority of the accomplishments being obtained by the minority of the employees. The vital few have the talent and ability necessary to generate ideas and carry them to fruition. These individuals provide the leadership to focus the attention and energy of the majority on the tasks required to produce the results, whether these results are new ways to present and promote products to the customers or they are new chemical formulae resulting in a totally different line of products. While it is true that sometimes new methods and products are produced by one of the many, this is the exception and often catapults that individual into the ranks of the few. In any large group it is approximately 20% that produce most of the results, but it is not always the same 20%. There is a flux from the many to the few and back again and only the rarest of individuals remains within the ranks of the few for a long time.

On the other hand, in any organization it is possible to identify a small group of individuals that produce most of the interruptions. In company town hall meetings the same individuals ask the questions that are later discussed over lunch; the questions that put the boss on the spot, or open the floor to discussions that upper management would just as soon have postponed until they arrived at some decision before making the problem public. It often seems that the 20% group producing 80% of the interruptions has a certain sensitivity to what situations cause the administration of the organization the greatest embarrassment. They seem to know that the new travel restriction are being violated by the company president; that the new local policy for sick leave is inconsistent with that in the corporate office; and they certainly know the most current office rumors.

The problems inhibiting the success of an organization often appear to be mutually independent, but in point of fact, they are not. Problems, like results, are interconnected, however subtly, so that generating solutions is very similar to generating results. What this means in practice is that certain solutions are connected to multiple problems within an organization; the distribution of solutions have the same imbalance in their influence on problems as individual talents have on the distribution of income among people. This imbalance is manifest in 80% of the problems being resolved by 20% of the solutions. In a complementary sense most solutions do not have a substantial impact on an organization's problems and their implementation are costly in the sense that the cost per problem resolution is approximately the same for all solutions generated. Consequently, it would be highly beneficial to determine if a given solution is one of the 20% before it is implemented and thereby markedly reduce the cost associated with problem solving.

As a final example, the majority of the activities of most people do not produce significant results, that is, results of the 20% kind; whether the activity is research, problem solving or even painting. There is a story regarding the artist Pablo Picasso, that may be apocryphal, but is no less relevant here. As you know, Picasso was a notoriously prolific artist and many copied his various styles. An acquaintance of his brought a painting to him and asked if he was the artist, to which Picasso replied: "It is a fake." Being unsure of this response the person went back to the art dealer from whom he had bought the painting. The dealer verified that the painting was, in fact, an original Picasso. The next day the person went back to the artist and confronted him with the claim of the art dealer, to which Picasso replied: "Yes, I painted it. I often paint fakes."

The point of this story is that Picasso realized that not every painting he did was a masterpiece, and if it was not a masterpiece, in his eyes it was a fake. Consequently, a significant fraction of Picasso's time was spent in producing fakes; perhaps not 80%, as it would be the normal artist, but some large fraction of his time. The same is true for mere mortals like us. If left unchecked, most of what we do is not significant and to change this requires a conscious effort of will.

We have argued that what is necessary to build a science of networks is an interdisciplinary approach to the study of reality, not confined to the physical sciences, but ranging from biology to psychology, neurophysiology and the study of brain function [125], and from there to the social sciences [47]. In particular, network science is not necessarily a consequence of reductionism. The clearest example of the weakness of reductionism in the physical sciences is given by the theory of phase transitions using renormalization group theory. This theory specifies a set of rules for establishing the critical coefficients of phase transition phenomena. Wilson and Kogurt [194] prove that the value of these coefficients can be assessed with theoretical models in a way that is totally independent of the detailed nature of elementary interactions among the elements in the phase transition network. In other words, the renormalization group approach establishes the independence, or at most weak dependence, of the different levels of reality, and, even if in principle a man is nothing more than a collection of atoms, his behavior has to be studied, with scientific paradigms which do not have anything to do with the dynamics of atoms.

4.3   What we know about complex networks

A complex network consists of a set of dynamical elements, whatever their origin, together with a defining set of interactive relationships among those elements. It is also possible to study a subset of elements, called a subnetwork of the network. A physician who is interested in the human body as a network may specialize in the brain and/or nervous system, as subnetworks. The network may interact with an observer, who may be a member of the network itself, or who may be part of the environment. A psychologist, for example, may argue with a physician over where the network called a person ends and the external environment begins. But we sidestep these interesting but tangential issues.

Let us refine the distinction between information and knowledge which, in turn, leads to two distinct aspects of complexity. On the one hand, the complexity of a phenomenon depends on the knowledge sought by the observer, and what constitutes this knowledge depends on the purpose of the investigation. Imagine that a network is studied to understand it, namely to describe and control it, or to predict its dynamic properties, but this is not always the case. For example, weather cannot be controlled, but it is very useful to make accurate short-term forecasts. Predicting the trajectory of a hurricane may save millions in dollars, not to mention the saving of lives, even if, in principle, we cannot know the hurricane's fundamental nature. On the other hand, complexity also has to do with intrinsic structure and such structure is measured by the information content of the network. This information is distinct from knowledge in that it is independent of the observer and the observer's theories, but we would relate the knowable with the free information of the last chapter and the unknowable with the bound information.

As an example let us examine a particular network, say a network of scientists. To determine how to measure the success of a collection of scientists we need to understand how such human networks operate. Suppose a large cohort group has a hypothetical set of choices, say these choices consist of a large set of nodes on a computer network to which they may connect. If we assume that the choices are made randomly, that is, independently of the quality of the nodes, or of the selections made by other members of the group, the resulting distribution in the number of times a given node is selected has the familiar bell-shape. In this case the selection process is completely uniform with no distinction based on personal taste, peer pressure or aesthetic judgment. In this way a random network of links between nodes or more generally connections between individuals within the network is made.

A bell-shaped distribution describes the probable number of links a given node has in a random network. Barabási [8] determined such distributions to be unrealistic for describing many common phenomena. For example, scientists were able to show that the number of connections between nodes in real-world computer networks deviate markedly from the bell-shaped distribution. In fact they found that complex networks have the expansive inverse power-law distribution of Pareto rather than the confining distribution of Gauss. Furthermore, scientists recognized that phenomena described by such inverse power laws do not possess a characteristic scale and referred to them collectively as *scale-free* networks, in keeping with the long history of such distributions in social networks [124].

The distribution of connections discussed above has a very specific interpretation if the network of interest consists of scientists. The way scientists connect to one another is through their published work, the presentations at conferences, symposia and workshops and by means of citations. If a scientist is doing an experiment s/he discusses at great length the limitations of the previous experiments that have been done and explains the theory that this new experiment is introduced to test. Throughout such discussions the investigator liberally cites the works of others so that the reader has a clear understanding of how the experiment being done fits into the larger body of work or how the calculation supports or rejects the experimental results of others. Most importantly the scientist establishes the significance of his/her work through the connectivity of that research to the research of others. Those papers in the research area that are considered seminal, which is to say, those that lay out the fundamental problems that must be resolved or which explain how one or more such problems may be explained, establish the context within which scientists place their own research. Consequently, such papers are referenced by nearly everyone in that area and it would be unreasonable to expect the distribution of connections to be bell-shaped. In fact, most measures of scientific productivity are found to be inverse power law [161].

We propose to discuss scientific research as a process of achievement so as to make the presentation somewhat less restrictive. We examine the distribution of the process of achievement and consider the mechanism to explain why such distributions have long tails, such as the one shown in Figure 2.5. A complex task or process, such as achievement, has been shown to be fundamentally multiplicative rather than additive because it requires the successful completion of a number of separate tasks, the failure of any one of which would lead to the failure of the overall task. We could be talking about achieving a given level of income as we did earlier; achieving a level of skill as a training objective, for example, in the military; or we may be interested in biological evolution or the interaction of physiological networks; but for the moment we restrict the discussion to social phenomenon. In particular, consider what a scientist does in order to publish a scientific paper. Remember that it is the process that is of interest here and not necessarily how publishing a paper is achieved so the format can be applied to any complex social network.

Part of being a good scientist is being able to identify a significant problem. In today's environment this usually means the ability to partition a difficult problem into manageable pieces that make interesting, but relatively short papers for publication in scientific journals. This strategy for diffusing scientific knowledge through the publication of a number of relatively short papers rather than a single tome is very different from the nineteenth century when scientists would work out complete solutions to fundamental problems and once obtained, publish a monograph containing all the details of the extended study. The scientists of that age were often financially supported by their families and could delay the publication of findings until the research was complete. The delay of publication was acceptable because science was not a job; it was a vocation. With the need for financial support, scientists today rarely take the long view, but more practically focus on the more readily attainable in publishing their work. Each publication addresses a question related to the long time research goal, but is more manageable in its scope, and over the course of time the separate publications form a mosaic that elucidates the larger research issue. The question of manageability for today's scientist is a matter of having the tools required to address the successive pieces of the general question selected for investigation. Of course one must be able to identify when a particular piece of the problem has been solved, or enough of the piece has been understood to make the result interesting and worthwhile. At this point a productive scientist stops and documents the results for publication, however stopping at this point and writing up what has been done is not a welcome task. For many investigators the desire to push on with the research is too strong and they fail to consolidate what they have learned and establish priority. Consequently, these scientists often find their results published in the trade journals obtained independently by more conscientious, less ardent and sometimes less talented colleagues.

One of the many reasons for not taking the time to write a paper is that most scientists have no idea how to write, or at least not how to write interestingly. The truth of this is so evident that when a scientist does come along who can both write well and do good science, s/he becomes a legend. But even when a scientist can write adequately, s/he is rarely comfortable in that role and often hurries back to the test tube, computer algorithm or blackboard; whatever their research comfort zone.

Of course communicating is only one factor in this process of explaining research results. Most scientific manuscripts are not accepted for publication the first time they are submitted to a journal. The procedure is for the editor of the journal to find one or more fellow scientists to act as reviewers to evaluate the manuscript using a number of criteria. The criteria are not universal, but usually contain such things as innovation of the research, the clarity of writing, and the reliability of the results. These are all characteristics that the journal requires a manuscript to have in order to publish it. Typically a referee rejects a first-time manuscript or accepts it for publication subject to a number of changes. The referee provides feedback on all aspects of the manuscript including suggestion on how to make the results more significant using statistical measures, possible reorganization of the arguments and the rearranging of material including shortening the introduction and adding appendices. In short, a referee has no limitation on what s/he may require before recommending the manuscript for publication. In the final analysis a referee's only guide is their own scientific perspective and personal understanding of what constitutes good science. Refereeing a manuscript is a very subjective enterprise.

The author must respond to all of the referee's criticisms to proceed to the next step in the publication process. Of course this does not mean a scientist timidly submits to the demands of the referee, even though most times the comments are made in good faith. The author must weigh, interpret and determine if the suggestions change the meaning of what s/he is trying to communicate. Most importantly, the investigator must be willing to profit from constructive criticism and exercise a certain amount of tact in responding to the referee's objections. It usually serves no purpose to point out that the referee does not understand the manuscript or that the suggestions do not make any sense. The referee's confusion may also be taken as an indication that the technical material is not clearly explained. Consequently, even the most obtuse evaluation of a manuscript can serve a useful purpose for an investigator determined to publish.

We can transform the above discussion on publishing a paper into quantitative form by associating a probability with each of the abilities discussed: the ability to dream up a good topic, the competency to do the necessary research, knowing when to stop and write up the work, sufficient humility to accept criticism, tenacity to make changes and resubmit the manuscript and so on. To some level of approximation the overall probability of publishing a paper is the product of these individual probabilities. The more complex the process the greater the number of probabilities contained in the product. The number of papers published by scientists over time depends on all these factors and many more. The rate of publication is therefore not a constant over time but changes in a random way due to the variability in the above factors. For example, for one paper a referee may be supportive and indicate how the investigator can improve the presentation. It is then relatively easy to follow the suggestions and publish the paper. In another instance the same referee may be acerbic and personally criticize the author in such a way that s/he withdraws the paper in a pique. The referee assigned to a paper is most often just a matter of luck in terms of their personal reaction, even though the editor may have given their selection a great deal of thought. Consequently the rate at which papers are published is given by the product of a random rate variable and the number of publications, both of which change over time.

Using the above argument the variability in scientific publication may be modeled as a random fluctuation in publication rate. The product form of the change in the number of publications and the publication rate tells us that if the number of publications is small the influence of a fluctuation is suppressed. On the other hand, if the number of publications is large the influence of a fluctuation is amplified. Therefore, the manner in which the publication network responds to changes in the environment depends on the state of the network. This behavior is typical of certain kinds of complex networks. The influence of the fluctuation on the number of papers to be published is determined by the number of papers already published. The more papers that have been published the more strongly the process responds to perturbations, both positive and negative. Some analysis of this situation gives us a distribution for the successful publication of a paper that is inverse power law. Such a distribution for the phenomenological measure of scientific productivity was first obtained by the biophysicist Lotka [98] and is depicted in Figure 3.5. Consequently, the distribution of the number of papers published by scientists, Lotka's measure of scientific productivity, is one of Pareto's children.

4.3.1 Random networks

The networks we make contact with on a regular and even daily basis are the power grid, the transportation network through roads trains and airplanes, or on a more personal level the neurons in our motor control network. For modeling purposes it does not matter what the nodes or elements of the network are. Given some spread of nodes we create a network by connecting the nodes as indicated by straight lines representing the nodal interactions, see Figure 4.1. Nodes that are not connected by a line have no direct influence on one another. Connecting airports by allocating restricted plane routes determines one of the country's major transportation grids. One could also examine the number of connections made with any particular node on the World Wide Web (WWW). How the connections are made between the nodes is all-important, since the connections determine the mutual interactions among the nodes and these connections determine what the network does, not just the properties of the individual elements. Different choices of connections can have devastating consequences regarding how successfully the network realizes its function. Of course we are also interested in the role played by architecture on the network's function.

The simplest complex network can be constructed by ignoring any network-specific design strategy that would distinguish between a computer network, a transportation network or a homeostatic feedback network within the body. One way to realize this goal is to randomize the connections made between nodes within the network as was done in the middle 1950s and early 1960s by the Hungarian mathematicians Paul Erdös and Alfréd Rényi, who investigated the properties of random networks and were able to draw a number of remarkable conclusions. Excellent accounts of the history of the Erdös-Rényi work on random graphs and the subsequent development of a theory of network connectivity is presented in a number of recent excellent popular books [8, 19, 180].

As connections are added to a random network, local clusters randomly form and these clusters are typically detached from one another. Consequently, the network as a whole is made up of a number of these detached clusters. However, as ever-more connections are added to the random network, eventually a dramatic change in the network character occurs, much like the transformation that occurs in water during freezing. Barabási [8] pointed out, that in this latter stage of the random network development it is possible to reach any node from any other node by simply following the links. Here every element interacts with every other element. Below the critical number of links the network consists of disjoint clusters of interacting nodes, whereas above the critical number the random network is a single immense cluster.

Random networks have a number of mathematically established properties. If the network is sufficiently large then all nodes have approximately the same number of links, however, there are certainly deviations from the average. In Figure 4.1 it is shown that some nodes have more links than the average and some have fewer links than the average. The distribution in the number of links to a given node in the random network of Erdös and Rényi is named after the French mathematician Poisson. The peak of the distribution is at the average number of links, just as it is for the Gauss distribution.

Recall how the properties of errors stimulated nineteenth century scientists and philosophers to interpret social phenomena in terms of average values. The same interpretation again emerged in the twentieth century from random network theory in which the theory predicts that most people have approximately the same number of friends and earn essentially the

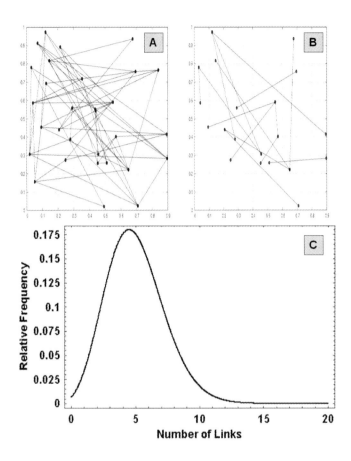

Figure 4.1: Typical random networks: [A] Average number of links to a node is 5. [B] Average number of links to a node node is 1. [C] Poisson distribution for the number of links to a node in a random network in [A]. A random network is obtained by randomly choosing pairs of nodes and connecting them. In [A] all nodes are connected, but in [B] the network is not fully connected.

same salary, most Websites have roughly the same number of visitors and most companies are of the same size. This lack of preferential treatment is certainly not characteristic of the world in which we live. Barabasi [8] and Watts [180] argue that this view on the importance of average values when applied to real-world networks, such as the WWW and the Internet has been shown to be wrong. In a complementary way West [187] has shown that the same is true in fractal physiology and medicine.

The small-world theory began as a theory of social interactions in which social ties can be separated into strong and weak. Clusters form for closely knit groups, just as in random networks, clusters in which everyone knows everyone else. These clusters are formed from strong ties, but then the clusters are coupled to one another through weak social contacts. The weak ties provide contact from within a cluster to the outside world, that is, to another small cluster. It is the weak ties that are all important, for interacting with the world at large.

The small-world phenomenon can be explained by the observation that the above clustering is retained if a few random connections are included. In this *small world* there are short cuts that allow for connections between one tightly clustered group to another tightly clustered group very far away. With relatively few of these long-range random connections it is possible to link any two randomly chosen individuals with a relatively short path. For example, it is far more likely that two friends of a given person know each other, than it is that any two individuals selected at random in the United States know each other. The clustering coefficient measures how much greater the likelihood is of two people knowing each other in a social network over what it would be in a random network as a whole. Consequently there are two basic elements necessary for a small-world model, clustering and random long-range connections, which can be interpreted in terms of the concepts of regularity and randomness. This should suggest the connection between the small-world model and the exotic statistics of Pareto.

These are the bare essentials of the Watts-Strogatz small-world model [169] in which a level of clustering is measured by a coefficient that determines how interconnected people are. A complete discussion and the further development of network theory is in the excellent account of this history by Watts [180].

Small world theory models the network nodes through the number of connections at each node and the strength of the connections between various nodes [192].[1] One property of such self-similar networks is that it appears to be random. However, like the network of airports some nodes become hubs having substantially more connections than does the *typical* airport; another manifestation of the Pareto Principle. Individuals are like that as well, with some having an order of magnitude more *friends* than others. This clustering of connections at certain nodes, as you may recall from the discussion of income, is one of the characteristics of inverse power-law distributions.

4.3.2   Scale-free networks and the STDs epidemic

The uncorrelated randomness of the Erdös-Rényi network does not exist in real world complex networks. This difference between the real world and an idealized mathematical model of the world is to be expected, in particular when the mathematical model is based on an assumption of pure randomness. It is precisely the difference between the real and simulated words that leads the scientist to those properties of most interest, that which is not explained has not been included in the model. The Poisson distribution, as was mentioned, describes the probable number of links a given node has in a random network. The distribution is based on the assumption that the probability of any new connection is independent of how many connections already exist at a node. The fact that real networks do not have Poisson statistics is strong evidence that the underlying assumption of independence is wrong. This is precisely the mathematical condition that Barabási and Albert [9] determined to be unrealistic, that is, this independence is not a property of networks in the real world. In fact they found that complex networks have the inverse power-law distributions discussed in the chapter on information. Furthermore, they recognized the scaling character of the inverse power-law distribution and the lack of scale in phenomena described by them. They referred to them as scale-free networks in keeping with the long history of such phenomena.

Of course inverse power-law phenomena are not new to the present discussion. As we have already discussed in the previous chapters they appeared

---

[1]For example, if k is the number of connections at a given node, it is found that complex networks, such as the Internet, have a Pareto-like inverse power-law probability density of the number of connections to a given node. This means that the probability that a given node has k connections is

$$P(k) \propto k^{-\alpha},$$

where α is a positive definite power-law index.

in the seminal work of Pareto in economics, the works of Zipf [195] on natural languages, the study of Auerbach on the size of cities, the investigation of Lotka on the number of scientific publications and in the research of Willis on bio-diversity, all verifying that complex systems are often inverse power law and by implication scale-free. What is new in the present discussion is the beginning of a mathematical language that may provide a way to discuss these things unambiguously. In particular, Barabási and Albert hypothesized two mechanisms leading to the inverse power law in the network context. One of the mechanisms abandons the notion of independence between successive connections within a network: the principle that the rich get richer. In a network context this principle implies that the node with the greater number of connections attracts new links more strongly than do nodes with fewer connections, thereby providing a mechanism by which a network can grow as new nodes are added.

The scale-free nature of complex networks affords a single conceptual picture spanning scales from those in the WWW to those within a biological cell. As more nodes are added to the network the number of links made to existing nodes depends on how many connections already exist. In this way the oldest nodes, those that have had the most time to establish links, grow preferentially. Thus, some elements in the network have substantially more connections than do the average, many more than predicted by any bell-shaped curve, either that of Poisson or Gauss. These are the nodes out in the tail of the distribution. In the transition from disorder to order, decreasing entropy or increasing information, networks shed their uncorrelated random behavior, characterized by average values and become scale-free, where the network is dominated by critical exponents.

After all is said and done, whenever a new connection is to be made, regardless of the kind of network, the principle of preferential attachment [8] comes into play. In deciding which book to read next, selecting among the titles, the fact that one of them is on the New York Times best sellers list could be a deciding factor. Preferential attachment is the added importance placed on one node over another because of prior decisions, usually made by others.

An interesting example of a scale-free network in the social world is the web of sexual contacts. It has been shown [88] that according to a data set of a few thousand individuals (ages 18-74 yr) gathered in a 1996 Swedish survey of sexual behavior [87] the cumulative distributions of the number of sexual partners decays as a power law for ranges larger than a given quantity.

During the twelve months prior to the survey it was found that the power-law index was $\alpha = 2.54 \pm 0.2$ where the number of sexual liaisons was in the range $k > 4$ for female, and $\alpha = 2.31 \pm 0.2$ in the range $k > 5$ for males. The cumulative distribution of sexual partners over all years since sexual initiation shows $\alpha = 2.01 \pm 0.3$ in the range $k_{tot} > 20$ for female, and $\alpha = 1.6 \pm 0.3$ in the range $20 < k_{tot} < 400$ for males. See Figure 4.2.

The inverse power-law distribution for sexual liaisons is quite interesting, in particular when compared with other kinds of social networks such as the distribution in the number of friends or the number of acquaintances in a Mormon community [2]. The latter networks show a probability distribution of links whose tail seems to fall off as a Gaussian function, suggesting that these networks are characterized by a single scale. Note that pure random networks are characterized by an exponential tail that decays slower than a Gaussian tail. Thus, a network with a Gaussian tail is characteristic of a web where multiple interactions are damped by some constraints.

So, what is the difference between the above two kind of networks? Probably, the inverse power law observed in the web of sexual contacts derives from a multiplicative stochastic process in which *the rich do get richer*. Probably, the skills and/or the lustful need for having new partners may increase in an uncontrolled and unconstrained manner as the number of previous partners grows. For example, as Figure 4.2B shows, the average number of partners since sexual initiation for males is 15 connections. Nonetheless, one male had 800 different partners. This is 50 times larger than the mean! The fact that the observed number of connections can be much larger than the average is what makes the web of sexual contacts scale-free, i.e., our intuition fails to establish a scale.

In the case of a web of friends the network looks regulated by additional mechanisms that not only prevent the activation of a multiplicative stochastic process, but even prevents the formation of new links once a certain number of links per node is reached. Thus, we denote the network as sub-random. One of these constraints is probably a requirement of *commitment*. In fact, social networks based on committed relationships such as friendship require a state of being bound emotionally or intellectually to another person or persons. These deep bonds can only form with time and requires some persistence and continuity. The effort to form such relationships can be directed toward only a limited number of people. Hence, the distribution appears to follow a Gaussian shape.

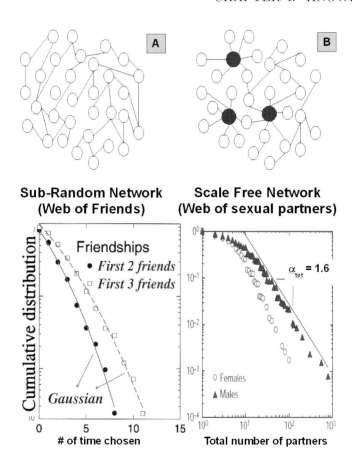

Sub-Random Network   **Scale Free Network**
(Web of Friends)   **(Web of sexual partners)**

Figure 4.2: [A] Example of sub-random networks such as a network of friends: each node has a few connections, multiple connections are damped. Linear-log plot shows the cumulative distribution of connectivities for the friendship network of 417 high school students, that is, the number of times a student is chosen by another student as one of his/her two (or three) best friends. The lines are Gaussian fits to the empirical distributions. [B] Example of a free-scale network: while most nodes have a few connections, there are some hubs with many connections. The figure shows the cumulative distributions of the number of sexual partners over all years since sexual initiation in a 1996 Swedish survey of sexual behavior [87]. Note that the distribution is linear in a log-log plot, indicating an inverse power-law behavior typical of scale-free networks. (The figures are adapted from [88] and [2].)

On the contrary, the web of sexual contacts appears to follow an inverse power law when these contacts depart from a committed relationship and become casual. No real commitment is required in this situation and adding new links, that is, having new sexual partners becomes inexpensive and simple. Under this condition an unconstrained multiplicative stochastic process nourished by different preferential attachment mechanisms, which make the process super-random, end up dominating the social intercourse.

This finding is, indeed, quite important. In fact, the scale-free structure of the web of sexual contacts indicates that it is sufficient that sexual transmitted diseases (STDs) such as HIV reach a few individuals with many sexual connections and it is subsequently easy to start an epidemic. Thus, for reducing the propagation of STDs it is necessary to develop education campaigns that strategically target, directly or indirectly, those individuals who may have large number of sexual partners: the hubs of the web of sexual contacts. If such individuals cease to function, it will lead to a failure of the web. It is possible to imagine two methods through which this goal can be accomplished: [a] by pinpointing hubs and neutralizing them in some way; [b] by targeting the specific mechanisms that nourish the multiplicative stochastic process that facilitates the connectivity of the web of sexual contacts, that is, adding constraints that prevent the nodes from connecting to the hubs. On the other hand, interventions that randomly target nodes, in this case individuals, may fail to break the web, in this case to reduce the propagation of the STDs.

The scale-free structure of the web of sexual contacts may indeed explain why sex education programs based on the mass distribution of condoms have generally failed to achieve acceptable results. For example, the failures in several African countries that have witnessed a rise of STDs infections concomitant with the application of these programs. In fact, in a scale-free structured web people with very few sexual contacts are still likely to be exposed to people who have had sex with many partners. It is actually possible to interconnect everyone in the network with a very few people. It is true that the network can be broken if someone identifies patients and promiscuous people and convince them to change their behavior and practice safe sex (or putting them in jail!): this is method [a] above. However, this is extremely difficult to achieve and it was thought that the mass distribution of condoms among the general population could solve the problem. Contrary to expectations, large-scale campaigns for promoting the random distribution of condoms only works among the population (because condoms have a given probability of failure), and a random attack on a scale-free network normally fails to produce catastrophic failure. This is one of the dominant characteristic of scale-free networks, their robustness to random failure (attacks).

On the contrary, a sex education program known as the ABC method has effectively succeeded in significantly reducing sexual disease infections. Uganda was the first country to develop and apply this revolutionary method and has experienced the greatest decline in HIV prevalence of any country in the world. Studies show that from 1991 to 2001, HIV infection rates in Uganda declined from about 15% to 5%. Among pregnant women in Kampala, the capital of Uganda, HIV prevalence dropped from a high of approximately 30% to 10% over the same period [167]. How did Uganda do it?

The acronym ABC stands for A=Abstinence, B=Be faithful and C=use Condoms only in the extreme cases. In other words the program focuses on strongly discouraging promiscuous sexual behavior and replaces it with a healthier life-style. Abstinence before marriage and marital fidelity are the most important factors in preventing the spread of HIV/AIDS and other STDs and have to be primarily advocated. Condoms do not play the primary role in reducing STDs transmission and are promoted only for those high-risk categories such as prostitutes and others who refuse to follow the two primary means.

The ABC program, indeed, has a better chance of efficiently reducing the spread of STDs by breaking the scale-free structure of the web of sexual contacts through the above method [b]. In fact, on one side the ABC method reduces the infectiousness of evident hubs such as prostitutes, on the other side with its primary emphasis on abstinence before marriage and marital fidelity greatly reduces the number of sexual links among the population. This has the effect of depleting the sexual hubs of their links and this ultimately causes a catastrophic failure of the web of sexual contacts through which the sexual transmitted diseases propagate. Note that policies based on condoms alone do not reduce the number of sexual contacts but may increase them through sexual licentiousness. At most condoms reduce the infectiousness of single sexual encounters, not the number of sexual partners per individual, which indeed increases! Thus, programs based only on mass distribution of condoms do not collapse the sexual web, at most they slow down the spread of the diseases in a *static* network. In reality the network is dynamic, it can increase or shrink according to the situations, and these policies could worsen the problem by making the network of sexual contacts through which STDs propagate more robust.

Returning to our discussion of scale-free networks, the above epidemiological example does not imply that the preferential attachment mechanism is always a conscious human act, since inverse power laws appear in networks where the nodes (people) have choice and in networks where the nodes (for example, chemicals) do not have choice, but are strictly governed by physical laws. But this mechanism implies that the independence assumption, necessary for constructing a random network, is not present in a scale-free network; one described by an inverse power-law distribution. In these latter networks not all nodes are equivalent. A different strategy leading to this result is given by Lindenberg and West [89], who show that such inverse power laws are the result of multiplicative fluctuations in the growth process as discussed earlier. Such multiplicative fluctuations are, in fact, equivalent to the mechanism of preferential attachment.

A random network lives in the forever now, whereas scale-free networks depend on the history of their development.

Why do scaling networks appear in such a variety of contexts; biological, chemical, economic, social, physical and physiological? In the case of the human lung, where the branching of the bronchial tree was shown to scale, it has been determined that a physiological fractal network was preadapted to errors, is unresponsive to random perturbations and therefore has a decided evolutionary advantage [184]. The evolutionary advantage would lead one to expect inverse power laws in all manner of biological phenomena, which we do in fact find. A decade later a similar result was found for scaling networks in that such networks, for example, the Internet, were shown to be fairly unresponsive to either random attacks or random failures [3]. An inverse power law in the number of links implies a large number of nodes with relatively few connections, but a non-negligible number of nodes could be considered to be hubs. Consequently, random attacks on such networks would most likely destroy or disrupt nodes with only a few connections and therefore have negligible effect. This would be one of the 80% and ignorable according to the Pareto Principle. The likelihood of striking a hub at random and thereby having a global effect is relatively small. Therefore the scale-free network is robust against random attacks. It is this tolerance to local failure that gives rise to the robustness of scale-free networks and this tolerance is a consequence of the architecture of the network. The scale-free network and fractal physiological structures both have an evolutionary advantage because of their fractal properties. On the other hand, such networks are susceptible to attacks directed at the 20%, the hubs, and elimination of only a small number of these hubs would produce catastrophic failure of the system. It is this later strategy that is adopted by terrorists, when these most sensitive elements of the network can be identified and attacked.

4.3.3   Scale-rich networks

The scale-free network formalism was constructed using a combination of data analysis, reasonable arguments concerning the interactions among individuals in a social network and mathematical modeling. The distribution of network connections resulting from this process was inverse power law in agreement with observation, as we have discussed. A completely different approach was taken by Carlson and Doyle [22], who wanted to know what kind of control process was necessary to design a network having the inverse power-law behavior. This is an engineering approach in which the constraints on the design of the network must be specified. Although both approaches result in inverse power-law distributions in agreement with data, other properties of the network are remarkably different, as we shall see.

The HOT procedure, discussed in the Introduction to this chapter, is intended to provide robust network performance in the face of environmental uncertainties. This engineering perspective forces power laws in specified network properties through tradeoffs between yield, cost of resources and tolerance of risks. As Carlson and Doyle emphasize:

> The characteristic features of HOT systems include: (1) high efficiency performance, and robustness to designed-for uncertainties, (2) hypersensitivity to design flaws and unanticipated perturbations, (3) non-generic, specialized, structured configurations, and (4) power laws.

Knowing the detailed properties of the individual elements of a complex network is of little or no value in predicting the behavior of large-scale interconnected networks. Physicists often look for universality in complex phenomena to explain scaling, whereas engineers look for network specific structures and environmental conditions incorporated into design to account for complexity. The self-organizing behavior of the Internet as new users and subnetworks are added to the network is one example of such adaptability.

We briefly sketch the arguments of HOT by defining the probability density $p(x)$ for an event occurring at the spatial location x and denoting the

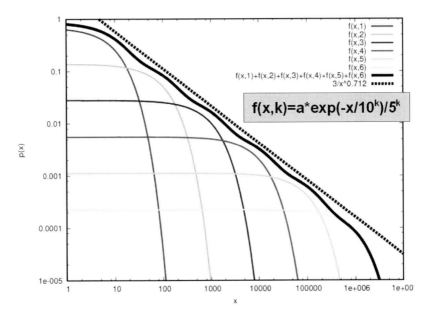

Figure 4.3: Example of a cumulative distribution of links per node for a scale rich network. In this case the network is the superposition of 6 different random sub-networks each characterized by a specific scale. Each of these sub-network is described by a exponential like distribution.

size of the region experiencing the event by the area $A(x)$. This is not unlike the argument put forward by Willis in 1922 in which he calculated the distribution of the number of species in a genera using the area of land involved. In an engineering framework there is a cost associated with the event, which they denote as $A^\alpha(x)$, where $\alpha > 0$ sets the relative weight of events of different sizes. But that is not the end of the story because we must allocate resources to cover the costs, or more properly to determine the size of the events. There are a number of ways to constrain the resources, the simplest being the total allocation of resources[2].

---

[2]Consequently, in the continuous case the expected cost of the avalanche of events is given by:

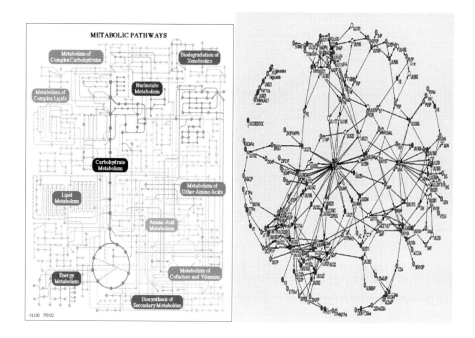

Figure 4.4: Examples of metabolic networks which are characterized by scale-rich structure.

$$\langle A^\alpha \rangle = \int p(\mathbf{x}) A^\alpha(\mathbf{x})\, d\mathbf{x} \qquad (4.1)$$

If $R(\mathbf{x})$ denotes the resource that restricts the sizes of the events and K is a constant, then the total quantity of resource is given by

$$\int R(\mathbf{x}) d\mathbf{x} = K \qquad (4.2)$$

The local event size is assumed to be inversely proportional to the local density or cost of the resource, so that $A(\mathbf{x}) = R^{-\beta}(\mathbf{x})$, where typically $\beta > 0$. Inserting this relation between the area and the resources into the cost function we obtain

$$\langle A^\alpha \rangle = \int p(\mathbf{x}) R^{-\alpha\beta}(\mathbf{x})\, d\mathbf{x}. \qquad (4.3)$$

The HOT state is obtained by minimizing the expected cost (4.3) subject to the constraint (4.2) and using the variational principle

Since its introduction the HOT technique has morphed into a number of related approached that have been applied to everything from the size of forest fires to the connectivity of the Internet. The modifications in HOT have to do with technology and economic design issues appropriate for the network under consideration, but these details are not pursued here. The point of interest to us is that one of the mathematical paths to be developed has to do with control theory and how to incorporate a variety of constraints into the design of a given complex network.

One version of HOT is Heuristically Optimal Topologies and is related to the design of the core and periphery of Internet's router-level topology [4]:

> ... a reasonably *good* design for an ISP network is one in which the core is constructed as a sparsely connected mesh of high-speed, low-connectivity routers which carry heavily aggregated traffic over high-bandwidth links. Accordingly, this mesh-like core is supported by a hierarchical tree-like structure at the edges whose purpose is to aggregate traffic through high connectivity.

Note the difference between the description of this network and the scale-free network of the previous section. Although both have the same node degree sequence, which is inverse power-law by construction, one is scale-free and the other is *scale-rich*. In the latter there is a distinction between being in a low-mesh core and the high-mesh periphery.

In other words the overall high variability and thus apparent power-law shown in the distribution of the number of links is created by a mixture or superposition of several networks centered in widely different scales. A very simple model is depicted in Figure 4.3. Here a superposition of a set of random networks each characterized by a cumulative distribution of links

$$\delta \left\{ \int p(\mathbf{x}) R^{-\alpha\beta}(\mathbf{x}) - \lambda R(\mathbf{x}) \right\} d\mathbf{x} = \mathbf{0} \tag{4.4}$$

Thus, the optimal relationship between the local probability density and constrained resource is given by the inverse power law

$$p(\mathbf{x}) \propto R^{\alpha\beta+1} \propto \frac{1}{A(\mathbf{x})^\gamma}$$

where the power-law index is given by $\gamma = \alpha + 1/\beta$. Consequently the greater resources are devoted to regions of high probability and we obtain an inverse power-law distribution.

per node described by an exponential curves is shown. The figure illustrates that if each network is characterized by a specific scale, their superposition may appear to present an inverse power law distribution. Of course, this is rigorously true in the specific case of the coefficients chosen in an opportune way. If the coefficients are chosen in a different way any kind of distribution may actually appear.

Scale-rich networks are easily found within metabolic networks where the high variability is due to the highly optimized and structured protocol that uses common carriers and precursor metabolites: see Figure 4.4. The inverse power laws are simply the natural null statistical hypothesis for such high variability data, rather than the traditional normal distribution.

The difference between scale-free and scale-rich is that in the former large fluctuations emerge and recede as a natural consequence of the internal dynamics. Consequently the network is robust. On the other hand, the latter network is hypersensitive to new perturbations that were not part of the design; so that the HOT network is robust, yet fragile. It should be emphasized that both types of networks have their Achilles' heel. How to identify these weaknesses and design around them is part of the challenge.

4.4   The Sun-climate linking: an ongoing debate

In Chapter 1 we briefly explained why the modern physics of complex phenomena should be considered as consisting of three components: experiment, computation and theory, which correspond to data, information and knowledge, respectively. This three-tiered structure allows equally valid alternative methodological paths for investigating a given phenomenon, which may yield to different interpretations. Under the philosophical assumption that a given phenomenon may have a unique explanation, the above alternative interpretations may be interpreted as just corresponding to alternative partial understanding of the phenomenon, which are relative to different but equally valid points of view. Evidently, until a phenomenon is fully explained, it is not possible to discriminate among its alternative interpretations, which are equally valid in that they are based on correct, although alternative, scientific methodologies.

In the previous chapter we briefly discussed how the above structuring of scientific problems apply to the climate change phenomenon. We mentioned the United Nations' IPCC's scientific interpretation of climate change. The IPCC report concludes that the contribution of solar variability to global warming is negligible and this finding has become the *majority opinion* thanks to the popular media, these being the television, radio, newspapers and computer blogs, all contending that the issue has been resolved and that the majority of scientists concur with the IPCC conclusions. Thus, it is reported that the *majority* of scientists believe the average warming observed since the beginning of the industrial era is due to the increase in anthropogenic greenhouse gas concentrations in the atmosphere.

What is not properly emphasized is that the IPCC findings are based on the analysis of global warming done using large-scale computer codes that try to incorporate all identified physical and chemical mechanisms into global circulation models in an attempt to recreate and understand the variability in the Earth's average temperature. The engineering of these models is not as straightforward as many believe but, as we explained above, is representative of specific methodological paths developed for investigating the phenomenon. The particular methodologies underlying the current climate models were adopted not because they were *known* to be the *correct* ones, but because they were *believed* to be the most *plausible* in conformity with a pre-understanding of the climate change phenomenon based on previous knowledge. So, it is important to analyze how this knowledge has progressed.

In fact, the intrinsic complexity of climate does allow for the possibility that the adopted methodologies are flawed. However, this has to be understood, not in the sense that these models are scientifically erroneous, but in the sense that they are representative of specific methodological strategies or paths that, although scientifically correct in principle, can be misleading in that they may lead to poor understanding of the phenomenon. This can be realized once the historical scientific environment in which the current climate models developed is made explicit. In the following subsections we briefly attempt to describe what in our opinion happened and how the current knowledge about global warming was formed. Then, we discuss an alternative explanation of climate change based on a phenomenological reconstruction of the solar effect on climate, which we developed, that suggests that the solar variability may have played a major role in climate change.

4.4.1   The Hockey Stick graph and the climate models

In Chapter 2 we discussed the importance of the data and in particular we briefly introduced the Hockey Stick temperature reconstruction by Mann *et al.* in 1999 [104], see Figure 2.20A. We explained how this secular temperature reconstruction has been interpreted as the most compelling evidence that the warming observed during the last century has been anomalous. In fact, from the Medieval Warm Period to the Little Ice Age the Hockey Stick temperature graph shows that the global temperature decreased by just 0.2 °C degrees against the almost 1 °C degrees warming observed during the last century.

Herein, we propose that the Hockey Stick temperature graph has either directly or indirectly conditioned the global warming debate. It has done so if only by giving *suggestions* about how the climate models ought to be written, that is, which mechanisms ought to be stressed and incorporated into the dynamic description, at least as a first approximation, and which ones could be neglected. Once the models were constructed, after several years of simulation, which have been used to interpret climate change, the results could only reflect the pre-understanding of the physics adopted in constructing the models themselves. In other words, the Hockey Stick graph suggested that the global warming observed during the last century was anthropogenically induced, climate models have been written *implicitly* incorporating this pre-knowledge. Evidently, once the models were completed, the computer climate simulations yielded results consistent with the information used to construct the models themselves. The conclusion that the global warming observed during the last century was anthropogenically induced was consequently built into the large-scale computer codes.

As the reader can see, the above situation is a subtle example of circular reasoning, and as such constitutes a logical fallacy. However, it is normal in science to have an idea (what we called a *pre-understanding*) about a specific phenomenon, prior to constructing a model and, finally, to check if the model works by comparing the theoretical predictions with data. The initial idea is indeed generated by some interpretation of the antecedent data from a given phenomenon, and the model ought to generate predictions consistent with the data used to construct it. The reason why science does not fall into the fallacy of circular reasoning is because Galilean physics requires that a scientific model be tested, not only against the antecedent data that inspired the model, but also against data from controlled lab experiments.

The physical constraints and conditions in the lab can be chosen by the experimenter to test the model in different situations. These new data are compared with the model predictions and it is the latter comparisons that are significant. Reproducing the data from which the model was build only establishes the lack of internal contradictions. The internal consistency of a model is certainly necessary, but it is not sufficient to determine the physical accuracy of the model.

Evidently, this experimental verification cannot be accomplished with climate. It is not possible, for example, to reproduce climate in a lab under alternate conditions; we cannot test a climate model for the case in which human emission during the last century are half their measured values. While we can run a computer simulation, we do not have data about this non-existent situation; nor can we reproduce climate in a lab under this alternate scenario; nor can we go back in time and see what would have happened if human emissions were cut by half during the last century. Thus, a climate model can be tested only against the same data that may have inspired it! This situation sets the stage for circular reasoning that may give the impression of *scientific consistency* just because model prediction and data appear consistent with each other.

There are at least two ways to test a model in the above situation: 1) compare the model prediction with data collected in the future; 2) compare the model prediction with the details of the data. The former case requires time, at least a few decades. For example, the climate models predict future climate scenarios for the 21st century, as seen in Figure 1.17, evidently we need to wait 100 years to test whether these predictions, and therefore the models generating them are correct. By the time the predictions are realized, more advanced climate models will be available and nobody will be interested in testing whether the current models are correct, except perhaps for didactic purpose. About the latter case we observe that climate simulations have been able to only approximately reproduce a multi-decade smooth trend in the data, as seen in Figure 1.16. However, the model simulations fail to reproduce the fluctuations, or more importantly, the patterns observed in the data, which are interpreted as internal variability or noise. Are these fluctuations merely noise or errors of measurement, or perhaps they are evidence that the current models are incorrect and/or poorly reproduce climate?

We discuss the above issue in the following subsection. In the present section, we explain why the Hockey Stick temperature graph or similar paleoclimate temperature reconstructions showing a very small preindustrial

temperature variability of about 0.2°C suggest that the global warming of
the last century may have been induced mostly by anthropogenic emissions.
In other words we investigate how scientific intuition about the causes of
global warming may have been forming. This intuition may have constituted
the environment that inspired the current climate models that can only mimic
the same intuition that inspired them.

Immediately after the publication of the Hockey Stick temperature graph
by Mann *et al.* in 1999 [104] a set of papers attempted to interpret this result
and arrived at similar conclusions. Let us see how. One of these papers with
the attractive title *Causes of Climate Change Over the Past 1000 Years* by
Thomas J. Crowley was published on the prestigious magazine Science in
2000 [36]. The abstract of Crowley's paper states:

> Recent reconstructions of Northern Hemisphere temperatures and
> climate forcing over the past 1000 years allow the warming of the
> 20th century to be placed within a historical context and various
> mechanisms of climate change to be tested... The combination of
> a unique level of temperature increase in the late 20th century and
> improved constraints on the role of natural variability provides further
> evidence that the greenhouse effect has already established itself
> above the level of natural variability in the climate system.

Let us analyze Crowley's seminal paper. Crowley used four different climate forcing components deduced from secular total solar irradiance reconstructions, volcano activity, greenhouse gas and aerosol records: see Figure
4.5. A linear upwelling/diffusion energy balance model was then used to calculate the temperature response to the above estimated forcing changes for
1000 years: Figure 4.6A. Finally, the computer model output including all
available forcing was compared with two quite similar paleoclimate temperature reconstructions. Figure 4.6B shows the comparison between Crowley's
simulation and the Hockey Stick temperature graph by Mann *et al.*.

About the adopted energy balance model (EBM), it is important to observe that its performance is quite similar to many other climate models, as
Crowley himself states [36]:

> The EBM is similar to that used in many IPCC assessments and
> has been validated against both the Wigley-Raper EBM) and two
> different coupled ocean-atmosphere general circulation model (GCM)

simulations. All forcing for the model runs were set to an equilibrium sensitivity of 2 °C for a doubling of CO_2. This is on the lower end of the IPCC range of 1.5 °C to 4.5 °C for a doubling of CO_2 and is slightly less than the IPCC "best guess" sensitivity of 2.5 °C.

Thus, Crowley's results are not expected to differ significantly from those obtained with the other climate models commonly used at that time.

As Figure 4.6B shows, there appears to be a quite good agreement between the EBM simulation and the Hockey Stick temperature graph by Mann and colleagues. This agreement suggests that Crowley's EBM model and the physics implemented in it is essentially correct. Ultimately, from the model simulation, as also evident in Figure 4.6A, one deduces that the observed global warming since 1900 is mostly induced by the GHG and aerosol anthropogenic component. Although the model fails to correctly reproduce several patterns in the temperature record observed since 1900, as Crowley acknowledges, the apparent high quality correspondence observed between the simulation and the 1000 years temperature reconstruction is indeed quite impressive. Crowley concludes his paper with the following statement:

> The very good agreement between models and data in the pre-anthropogenic interval also enhances confidence in the overall ability of climate models to simulate temperature variability on the largest scales.

Thus, as the title of Crowley's paper states, the causes of climate change over the past 1000 years were understood. Indeed, not all patterns could be reproduced, but this may be considered irrelevant, after all just a simple climate model was used and the data reconstructions were known to have some error. Consequently, the results found by Crowley could be considered approximately correct, and it would have been unrealistic to think that they could be misleading. The high quality of the agreement between computer simulation and temperature reconstruction suggested that the physics of climate implemented in the climate models was sufficiently accurate and complete, at least to a first approximation. Papers like those of Crowley have constituted the basis of the *knowledge* about climate change that the majority of the scientific community had since 2000 and the conclusion was that most of the warming observed since 1900 was anthropogenic.

More precisely, as Figure 4.6A shows, since 1900 the sun could have contributed just 0.12 °C on an observed global warming of approximately 0.8 °C:

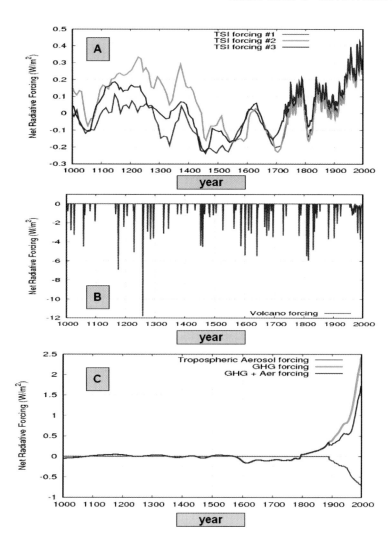

Figure 4.5: Forcing time series used in the EBM model runs by Crowley [36] (note scale changes for different panels). [A] Net climate forcing deduced from three different TSI proxy reconstructions that make use of ^{14}C ^{10}Be and sunspot number records. [B] Net climate forcing deduced from volcano eruptions records. [C] Net climate forcing deduced from the GHG and tropospheric aerosols records, the blue curve is the sum of the two components.

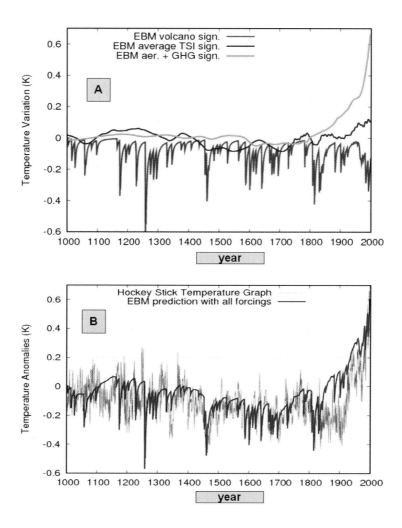

Figure 4.6: [A] Crowley's EBM model temperature output relative to each forcings shown in Figure 4.5. The green curve represents the EBM climate response to the GHG and tropospheric aerosol forcings; the blue represents the average EBM climate response to the three TSI forcings; the red curve represents the EBM climate response to the volcano eruptions. [B] Comparison between the Crowley's EBM model temperature output obtained with all forcings and the Hockey Stick temperature graph by Mann *et al.* [104].

thus, only 15% of the global warming observed since 1900 could be associated with increased solar activity. The other 85% had to be interpreted as due to the combined effect of volcano eruptions, tropospheric aerosol and GHG concentration changes. Because both volcano eruptions and tropospheric aerosol forcing have a cooling effect, the observed warming had to be associated with the anthropogenic increase of GHG in the atmosphere.[3]

It is evident that a strong interconnection exists between the Hockey Stick temperature graph by Mann and colleagues, the climate models, their interpretation and the knowledge we derive from it. As Crowley stated in the above quote the Hockey Stick temperature graph and its agreement with EBM simulations have "enhanced confidence in the overall ability of climate models to simulate temperature variability on the largest scales." This same confidence is the one underlying the IPCC reports. But what if the Hockey Stick temperature graph by Mann and colleagues is wrong and the climate experienced a much larger variability during the last 1000 years?

In Figure 2.20B and the related section we showed that since 2002 the Hockey Stick temperature graph has been seriously questioned. New paleoclimate temperature reconstructions, such as the one by Moberg *et al.* [113], present a significant pre-industrial climate variation (up to approximately a 0.65 °C variation from the Medieval Warm Period to the Little Ice Age). If Crowley's EBM simulation is compared with these more recent reconstructions the good agreement he obtained is lost, as Figure 4.7A shows. This comparison logically implies that if the secular climate during the last millennium experienced the large temperature variation suggested by the above more recent paleoclimate temperature reconstructions the EBM model adopted by Crowley is seriously incomplete, and by paraphrasing Crowley's quote itself we could state that this finding would *lessen* the "confidence in the overall ability of climate models to simulate temperature variability on the largest scales."

Indeed, it is possible to evaluate how Crowley's model should be corrected in such a way that its prediction is in better agreement with a temperature sequences such as Moberg's. With a simple multilinear regression algorithm[4]

---

[3]Note that Crowley's study was based on the TSI reconstruction by Lean in 2000 as we show in Figure 2.14C. If a more recent TSI reconstruction by Lean was used the solar contribution to the observed warming since 1900 would have been approximately 7.5% rather than 15%.

[4]A multiple linear regression model assumes that each contribution to the temperature change is linear with a coefficient adjusted to best fit the data.

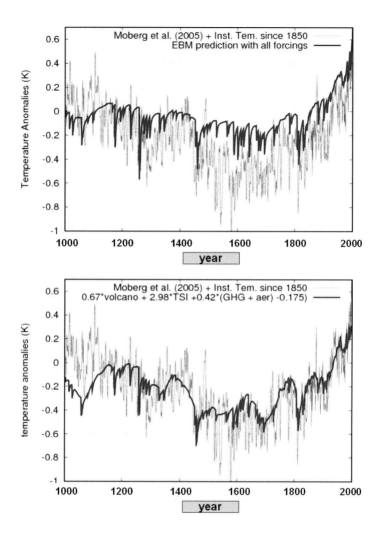

Figure 4.7: [A] Crowley's EBM model temperature output against Moberg *et al.* [113] palaeoclimate temperature reconstruction. [B] The Crowley's EBM model temperature output is adapted to fit the Moberg's reconstruction. Note that this fit requires that the volcano and GHG plus aerosol effect on climate must be significantly reduced by a factor of 1/3 and 1/2 respectively, while the solar effect must be increased by a factor of 3.

it is possible to calculate the factors by which each component of the simulation, corresponding to a given forcing, should be multiplied. Our best estimate is shown in Figure 4.7B where good agreement is observed between the corrected computer simulation and the paleoclimate temperature reconstruction by Moberg and colleagues.

We found that if Moberg's temperature reconstruction is correct, Crowley's EBM significantly overestimates both the volcano eruption effect and the combined GHG + tropospheric aerosol effect. The former must be reduced by 33% and the latter must be reduced by 57%. This is quite important because Crowley was already using a climate sensitivity to GHG change on the lower end of the IPCC range of 1.5 °C to 4.5 °C for a doubling of CO_2 and that was slightly less than the IPCC "best guess" sensitivity of 2.5 °C. Thus, the IPCC range of 1.5 °C to 4.5 °C for a doubling of CO_2 would be quite an overestimate. On the contrary, the solar effect on climate should be approximately 3 times stronger than what Crowley's EBM assumed.

Thus, it is evident that if the secular climate during the last millennium experienced the large temperature variation suggested by Moberg and colleagues, the anthropogenic effect on climate is significantly overestimated by the models, while the capacity of the sun to alter climate is significantly underestimated. In this scenario the sun may have contributed approximately +0.36 °C from 1900 to 2000 of the +0.8 °C observed, that is, approximately 45% of the observed warming during the same period. There is no need to add that this would have been much more than any estimates reported by the IPCC reports.

In summary it is evident that, by paraphrasing Crowley's quote, our "confidence in the overall ability of climate models to simulate temperature variability on the largest scales" does depend on the correctness of the paleoclimate temperature reconstructions of the last 1000 years. Our confidence about the correctness of the climate models would be *enhanced* if the climate experienced very little preindustrial variability such as the Hockey Stick temperature graph by Mann and colleagues shows, but, on the contrary, the same confidence would be seriously *lessened* if climate experienced a large preindustrial variability such as Moberg's temperature reconstruction shows.

The above reasoning draws us to the question: which paleoclimate temperature reconstruction is the most likely?

4.4.2 The 11-year solar cycle signature on climate: models vs. data

Part of the intellectual heritage of western science is logic, beginning with Aristotle's binary choice of A or not-A, the syllogism and ending with logical fallacies, such as the circular argument or circular reasoning. This last fallacy concerns us here because the argument has been made that due to the fact that some climate model simulations fit the Hockey Stick temperature graph and not that of Moberg, that the former is the more accurate temperature reconstruction. The circular nature of the argument is not immediately evident because the assumption is made that the climate model is correct, but this assumption is only justified by the agreement between model simulation and the Hockey Stick temperature graph of Mann and colleagues. Thus, which paleoclimate temperature reconstruction is most correct cannot be answered using simulation alone. To repeat: by accepting the above reasoning one would essentially say that those EBMs are correct because they agree with the Hockey Stick temperature graph and, consequently, that the Hockey Stick temperature graph is correct because it agrees with the same EBMs; an evident case of circular reasoning!

Because significant uncertainty remains regarding the paleoclimate temperature reconstructions, we need to address the problem from an alternative perspective. One idea is to identify a specific pattern in the temperature data that can be reasonably associated with a solar effect on climate and use this pattern for evaluating a phenomenological climate sensitivity to solar change.

It is evident that the most certain pattern characterizing solar activity is the *11-year solar cycle*. Neither the period nor the amplitude of the 11-year solar cycle is rigorously constant. The average duration of the sunspot cycle has been 11.1 years since the end of the Maunder Minimum in the 1700s, but cycles as short as 9 years and as long as 14 years have also been observed. Although the 11-year solar cycle was first observed in the sunspot count, all known solar variables follow the same 11-year solar cycle. In particular, as direct measurements of TSI from satellites have proven (see Figure 2.13), the TSI is observed to vary in phase with the 11-year solar cycle, with oscillations around an average of approximately 1366 W/m^2 showing a peak-to-trough amplitude from minima to maxima of approximately 1 to 1.5 W/m^2, and with fluctuations about the means of about $\pm$ 1 W/m^2 on time scales of a few weeks or months.

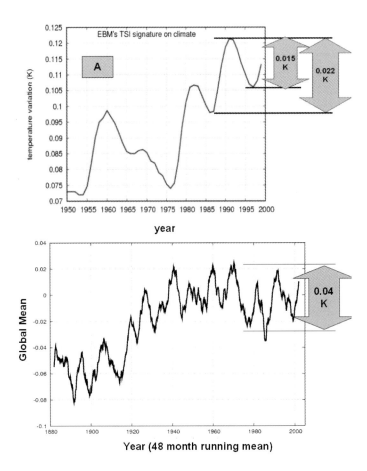

Figure 4.8: [A] Climate signature of the Total Solar Irradiance predicted by the linear upwelling/diffusion energy balance model by Crowley [36]. The curve is a magnification of the blue curve in Figure 4.6A. The peak-to-trough amplitude from minima to maxima of the 11-year solar signature is estimated to be approximately 0.015-0.022 ^{o}C since 1980. [B] The same prediction obtained with the 1880-2100 Climate Simulations with the GISS modelE coupled atmosphere-ocean model [62]. The global peak-to-trough amplitude from minima to maxima of the 11-year solar signature on the surface is estimated to be about $0.04K$.

It is claimed that this minimum-to-maximum TSI variation, which is approximately 0.1% of the TSI average level, is far too small to affect the Earth's climate. However this conclusion is directly based on the EBMs estimates. Figure 4.8 magnifies Figure 4.6A and shows the estimates of the global climate signature left by the TSI as obtained with the linear upwelling/diffusion energy balance model by Crowley [36]. The figure shows that the peak-to-trough amplitude from minima to maxima of the 11-year solar cycle signature on climate is predicted by the EBMs to be at least 0.015 to 0.022 °C since 1980. By using a different EBM, for example, the MAGICC climate model using best estimate parameters North, Foukal and Wigley [51] found that the 11-year solar cycle leaves on climate a peak-to-trough signature less than 0.03 °C (see the plot in their figure). So, typical EBMs' predictions regarding the climate signature left by the 11-year solar cycle are about 0.02 °C; indeed a very small effect.

The problem with the above theoretical finding is that it is incompatible with several other studies that have evaluated the 11-year solar cycle signature on climate by evaluating it directly from the global temperature record. Curiously, the IPCC report [162] is very explicit in these findings, as the following quote states:

> A number of independent analyses have identified tropospheric changes that appear to be associated with the solar cycle (van Loon and Shea, 2000; Gleisner and Thejll, 2003; Haigh, 2003; White et al., 2003; Coughlin and Tung, 2004; Labitzke, 2004; Crooks and Gray, 2005), suggesting an overall warmer and moister troposphere during solar maximum. The peak-to-trough amplitude of the response to the solar cycle globally is estimated to be approximately 0.1 °C near the surface. (From the IPCC Fourth Assessment Report, Working Group I Report "The Physical Science Basis," Chapter 9: *Understanding and Attributing Climate Change* pag. 674 [162].)

Thus, several phenomenological studies have estimated the peak-to-trough amplitude of the response to the solar cycle globally to be approximately 0.1 °C near the surface. This is up to five times the model predictions shown in Figure 4.8!

The above result can be found by using different methodologies. For example it is possible to filter the global temperature sequences and the solar sequences with appropriate band-pass filters, as shown in Figure 4.9

[145]. As the figure shows the solar Schwabe 11-year cycle leaves a peak-to-trough amplitude response of about 0.1 °C near the surface while the solar Hale 22-year cycle leaves a peak-to-trough amplitude response of about 0.06 °C. Note the good correspondence between the solar and temperature cycles.

Alternative ways to obtain the same result attempt to reconstruct the temperature sequence by means of a multilinear regression analysis that uses several known forcing such as the TSI, volcano, and El-Nino signals plus linear trends as done by Douglass and Clader [41] and Lean. At higher altitudes the 11-year solar signature on climate is larger [86]. Svensmarky and Friis-Christensen [171] have recently shown that there exist an evident correlation between the solar cycle and the negative of global mean tropospheric temperatures with galactic cosmic rays again after removing El Nino, the North Atlantic Oscillation, volcanic aerosols, and a linear trend: see Figure 4.11A.

Figure 4.10 shows a comparison between the empirical estimate of the peak-to-trough amplitude of the 11-year solar signature in the troposphere (from the surface to an altitude of 10 Km) obtained by Gleisner and Thejll [58] in 2003, and the correspondent global climate simulation estimates obtained by the GISS group [62]. The figure clearly shows that the model is significantly underestimating the solar effect on climate in the entire troposphere by a large factor between 3 and 8.

More recently, Camp and Tung [21] project the surface temperature data (1959-2004) onto the spatial structure obtained from the composite mean difference between solar maximum and solar minimum years and obtain a global warming signal of almost 0.2 °C that may be attributable to the 11-year solar cycle, which is even larger than 0.1 °C: see Figure 4.11B.

Thus, even if we may disregard the finding of Camp and Tung and hold the more conservative estimate of a peak-to-trough amplitude of the response to the solar cycle globally is estimated to be approximately 0.1 °C near the surface, we conclude that there exists a significant discrepancy between the climate model predictions and the phenomenological findings derived from the direct analysis of the data. Evidently, although it may be possible to argue that the data or the phenomenological studies may be erroneous and misleading, it is more probable that the climate models are *defective*. The next subsection is devoted to reconstructing the solar signature on climate by using the phenomenological findings derived from the data analysis itself.

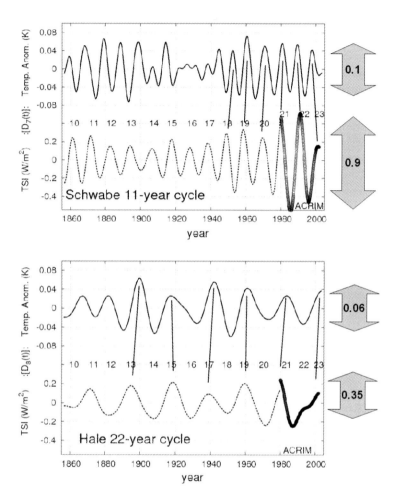

Figure 4.9: Multiresolution wavelet analysis of the global surface temperature and TSI sequences. [A] The band-pass filter emphasizes the variability around an 11-year periodicity. [B] The band-pass filter emphasizes the variability around an 22-year periodicity. The solar Schwabe 11-year cycle leaves a peak-to-trough amplitude response of about 0.1 °C near the surface. The solar Hale 22-year cycle leaves a peak-to-trough amplitude response of about 0.06 °C near the surface.

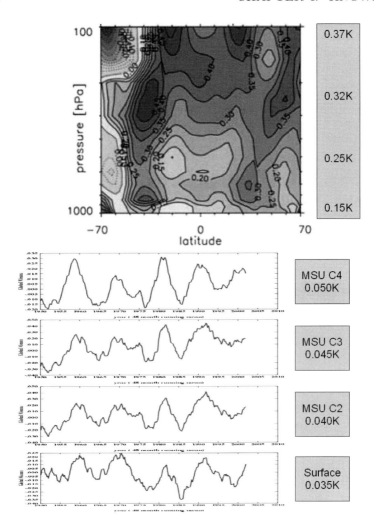

Figure 4.10: Top: empirical estimate of the peak-to-trough amplitude of the 11-year solar signature in the troposphere from the surface (1000 hPa) to about 16 Km of altitude (100 hPa) by Gleisner and Thejll [58]. The right panel reports approximate estimates of the global average amplitude of the signature. Bottom: The four panels refers to the GISS ModelE global average estimates of the solar forcing at different altitudes from the surface to the top troposphere (MSU channel 4). The small four right panels reports approximate estimates of the global average amplitude of the signature.

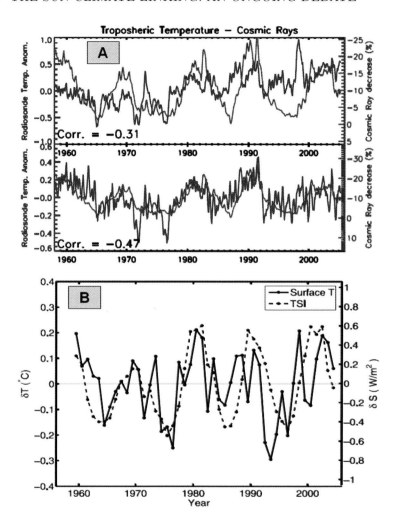

Figure 4.11: [A] The solar cycle and the negative correlation of global mean tropospheric temperatures with galactic cosmic rays. The upper panel shows observations of temperatures (blue) and cosmic rays (red). The lower panel shows the match achieved by removing El Nin o, the North Atlantic Oscillation, volcanic aerosols, and also a linear trend : adapted from Ref. [171]. [B] Comparison between the global-mean surface air temperature (1959-2004) (solid line, with scale on the left axis) and the annual-mean TSI time series (dashed line, with scale on the right axis): adapted from Ref. [21].

4.4.3   The phenomenological reconstruction of the solar signature on climate

The philosophical assumption on which a solar signature of climate can be reconstructed empirically or *phenomenologically* may appear foreign to the traditional climate model approach that requires the determination and the modeling of all physical mechanisms involved in a specific phenomenon. Thus, the usual critique that is formulated is that the empirical findings should be dismissed because no physical cause is explained nor determined: sunspot and temperature correlation does not prove causality it is stated again and again. Moreover, how can empirical studies be considered more scientifically accurate than numerous rigorous climate model studies that have appropriately taken into account all known climate physics and that essentially agree with each other?

Consider the first observation in which an apparent correlation between two variables does not *prove* a causal relationship: this is generally correct. However, we ask the weaker question of whether it is legitimate to conclude that a correlation between the solar record and the Earth's temperature *suggests* causality. Isn't the above critique subtly biased because in science causality is never *proven* but only *suggested* by some form of correlation between a model prediction and the data? Finally, what should we think about the traditional climate model approach when its predictions about the solar signature on the Earth's surface temperature appears to be less than 35% of what is observed in the data? While it is true that a correlation does not prove causality, doesn't a *lack* of correlation prove that a scientific model is likely to be erroneous or incomplete?

About the second observation concerning climate being a complex phenomenon. One of the characteristics of the science of complex phenomena and in general of any phenomena is that a physical model attempting to reproduce its characteristics must necessarily be based upon the current understanding of how the network under study really works. If there are still important aspects of a complex phenomenon that are poorly understood by the scientific community, this lack of understanding effects all modelers so they will tend to make similar modeling errors. Thus, especially when addressing a complex network, the apparently surreal but, indeed, very realistic scenario is that many scientific groups arrive at similar conclusions about a given phenomenon because they are using similar models that reflect the same incomplete scientific knowledge, and all of them may be wrong!

Only an improved understanding of the physics of a complex networks can eventually suggest to scientists how to construct more accurate physical models. Thus, the fact that several climate models do suggest that the recent global warming has been anthropogenically induced does not necessarily imply that this current theoretical understanding of climate change is correct. Not acknowledging this potential problem gives rise to a form of logical fallacy where a thesis is deemed correct only because it has a *tradition* or, as it is commonly stated in the media, has a *consensus*, and consequently discussion should cease and something should be done. However, history has proven again and again that science is based on careful interpretation of data, not on *scientific consensus.*

An empirical or phenomenological reconstruction of the solar signature on climate surely has its shortcomings and imprecisions. However, phenomenology is not supposed to be a substitute for a more traditional scientific methodology. Phenomenological reconstructions have the purpose of keeping alive the possibility that a given phenomenon may be addressed and explained from alternative points of view, and new mechanisms have to be sought and investigated. While computational studies stress the a correspondence between theoretical predictions and data patterns, empirical studies stress the discrepancies between predictions and data, and indicate that these differences suggest alternative explanations.

After all, it was the empirical astronomical findings of Galileo and Kepler with their limitations, controversies and sometimes misleading understanding (Galileo thought that the tides proved the Earth moved around the Sun) that challenged the established Aristotelian understanding of the cosmos and paved the way to a new scientific understanding of the world. This could occur only several decades later, starting with Newton. Analogously, it was the empirical findings of Michelson and Morley that suggested the non-existence of the ether and made possible Einstein's elegant solution summarized in his special theory of relativity. Empirical or phenomenological studies always come first, the hard science comes later.

The phenomenological approach to interpreting climate change assumes that the Earth's climate is forced by extraterrestrial mechanisms whose temporal variability is qualitatively reproduced by the TSI records, much as those shown in Figures 2.13 and 2.14. The physical nature of the actual extraterrestrial forcing of climate as well as the actual physical climatic mechanisms processing the external input remain incompletely understood.

The climate is assumed to be influenced by TSI forcing, as adopted by the current climate models [162] (see also Figure 1.19). But climate may also be influenced by cosmic rays that are modulated by the solar magnetic field and that may modulate cloud cover patterns [73]. Climate may also be more sensitive to specific frequencies in the spectrum than to others. For example, UV radiation can change the atmosphere chemistry by altering, for example, the stratospheric ozone concentration [123]. Both these solar-climate link mechanisms are absent from the computational climate models because their physics is not yet well understood. However, the interesting finding is that geomagnetic activity, cosmic ray flux and UV irradiance records are approximately mimicked in their decadal and multidecadal patterns by the TSI records.

There may also be many other astronomical-climate link mechanisms that are not well understood nor yet identified and, therefore, they remain a mystery. In any case, it is likely that any astronomical forcing, except a possible lunar effect on climate, are somehow linked to the TSI and qualitatively share similar patterns with it. This is possible for at least two reasons: 1) they are generated by the solar variability itself or 2) they share a common origin with solar variability. Consequently, TSI can still be considered a qualitative proxy of most astronomical forcing of climate, both those that are known and those that are still unknown.

Thus, when the TSI record is used as a proxy of the extraterrestrial forcing of climate it is intended that, for example, because TSI records show evident decadal cycles (the so-called 11-year solar cycle) we assume that the overall astronomical forcing of climate present similar cycles. Analogously, because the total solar irradiance records show a multidecadal complex up and down pattern that includes the Maunder minimum in the 17th century during the little ice age and since then an increase up to the modern maximum, we also assume that the overall astronomical forcing of climate presents a similar up and down pattern with a minimum during the 17th century and a maximum during the last decade.

On the other hand, as shown in the previous section, according to several independent studies focusing on climate data analysis, climate shows a peak-to-trough amplitude in the response to the 11-year solar cycle that is estimated to be approximately 0.1 °C near the Earth's surface. What is causing this signature? We have seen that according to the climate models the TSI forcing alone is not capable of explaining the magnitude of the effect (see Figure 4.8). Thus, if these temperature cycles are somehow linked to the 11-year solar cycles that needs to be established. Alternatively, if the models are incorrectly representing the climate response to TSI forcing, they are missing alternative solar-climate link mechanisms that could explain the observed discrepancy. In both cases, the actual climate models are incapable of correctly evaluating reproducing the climate patterns and are, therefore, incomplete or erroneous.

The phenomenological interpretation of this modeling failure is that although we do not know the mechanisms involved in the solar-climate link, the solar variability is nevertheless, directly or indirectly, associated with these decadal cycles in the Earth's near surface temperature. Thus, by using the TSI record as a proxy for the overall solar forcing on climate it is possible to measure a phenomenological sensitivity of climate to the 11-year solar cycle and this sensitivity is about $0.11 \pm 0.02 \ K/Wm^{-2}$ as found by Douglass and Clader [41] and Scafetta and West [145]. Thus, according to this interpretation an 11-year solar oscillation of about $1 \ W/m^2$ peak-to-trough amplitude would be the phenomenological "cause" of a temperature cycle with a peak-to-trough amplitude of about $0.1K$.

However, this is not enough to reconstruct a solar signature on climate with temporal scales smaller or larger than the decadal one. According to climate science, the climate network, on a large scale, is a thermodynamic network that responds to an external energetic input with a given characteristic time response. Everybody knows that the temperature of the water inside an ordinary kettle over the flame of a stove increases slowly until thermal equilibrium between the water and the heater is reached. Similarly, the climate network responds slowly to an external forcing, that is, an energetic input, trying to reach thermal equilibrium. Because the forcing continuously changes in time, the climate is always changing and evolving and thermal equilibrium is probably never reached.

The simplest way of modeling climate is by assuming that it responds analogously to an external forcing as a resistance-capacitor (RC) network responds to a change of voltage. As is well known, an RC network responds to the stimulus with a given relaxation time response, $\tau = 1/RC$. This model implies that the sensitivity of the network depends on the frequency of the input. The larger the frequency dampener, the larger the amplitude of the output signal for equal input signals. Using this model, climate sensitivity should decrease more at higher frequencies than it does at lower ones. Any frequency sensitivity can be easily calculated once the equilibrium sensitivity is evaluated.

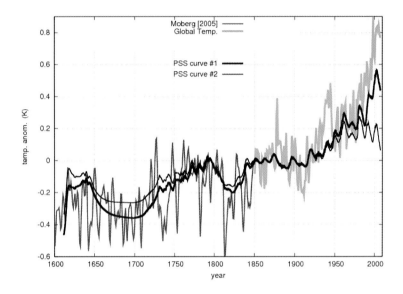

Figure 4.12: Phenomenological solar signature against a paleoclimate temperature reconstruction [Moberg *et al.*, 2005] from 1600 to 1850 (thin grey line) and global surface temperature record since 1950 [Brohan *et al.*, 2006] (thick black line). The PSS (thick grey) curve is obtained with $\tau_1 = 0.775$ year, $\tau_2 = 12$ year and $k_{1S} = 0.052 K/Wm^{-2}$ $k_{2S} = 0.44 K/Wm^{-2}$ with the model forced with the TSI reconstruction [A]. The PSS (thin black) curve is obtained with $\tau_1 = 0.775$ year, $\tau_2 = 8$ year and $k_{1S} = 0.052 K/Wm^{-2}$ $k_{2S} = 0.30 K/Wm^{-2}$ with the model forced with the TSI reconstruction [C]. [from [150] with permission]

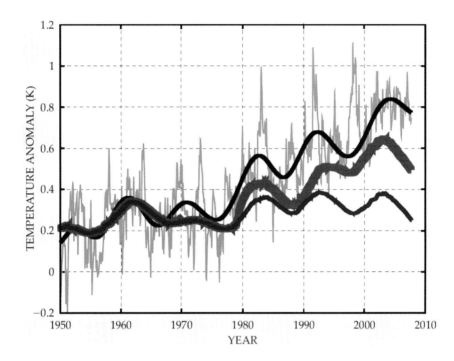

Figure 4.13: Global surface temperature anomaly (green). The green curve is the difference between the measured GST and the time average of GST between 1890 and 1910; it emphasizes the observed warming since 1900. The GST anomaly data are low-pass filtered to remove any volcanic signal and smoothed to stress the 11-year modulation (black curve). Two alternative Phenomenological constructions are shown that use a TSI proxy reconstruction until 1978 and two alternative TSI satellite composites after 1978: the ACRIM (red) and the PMOD (blue) [from [147] with permission].

Thus, a simple way to proceed is by evaluating the characteristic time response of the climate to perturbations. This response time can be obtained from the autocorrelation properties of the temperature record by using the fluctuation-dissipation theorem. The fluctuation-dissipation theorem states that the response of a system in thermodynamic equilibrium to a small applied force is the same as its response to a spontaneous fluctuation. Therefore, there is a direct relation between the fluctuation properties of the thermodynamic system and its linear response properties. The linear response can take the form of one or more exponential decays.

It has been found [148, 172] that climate is characterized by multiple characteristic time responses. In particular, the climate appears to be characterized by two major time scales: a fast time response of a few months, about half a year, and a decadal response with $\tau = 8$-12 years. This two scale response is reasonable because climate is the interaction of multiple phenomena. Some climate phenomena are characterized by a fast time response (just a few months) such as atmosphere phenomena. Others, such as the energy exchange between the atmosphere and deep ocean; the albedo changes due to the melting of glaciers and to the vegetation cover changes, are very slow processes that occur on a decadal time scale.

By using an RC-like phenomenological model that takes into account the above empirical findings (the 11-year signature and the two characteristic time responses), and under the assumption that the TSI record can be used as a proxy for all solar-related extraterrestrial forcing of climate, a secular phenomenological climate signature of the solar-related forcing can be obtained and is plotted against a secular temperature record: see Figure 4.12.

The phenomenological model reproduces most of the climate change patterns observed for the four centuries since 1600: The Little Ice Age during the 17th century which occurred in concomitance of the Solar Maunder Minimum; the global cooling during the Dalton Minimum in the early 19th; and the warming since then. This pattern of successive warming and cooling are clearly visible in the reconstruction. Figure 4.13 shows a rapid temperature increase from 1950 to 2008 that stresses the 11-year solar signature on climate and compares favorably with the model reconstruction. The correspondence of the observed and reconstructed 11-year cycles is evident and also the cooling observed since 2002 is explained as a consequence of the decreased solar activity.

The close correspondence between the temperature data and the phenomenological model reconstruction suggests that most of the observed climate change observed since 1600 has been caused by solar variations. Because it has been recently found [149] that the TSI increased since 1980, as the ACRIM TSI satellite composite does , also the observed global warming since 1970 has been induced mostly by solar variations, up to two-thirds of the trend (red curve in Figure 4.13).

This result greatly differs from the IPCC estimates according to which the sun is supposed to have caused less than 10% of the warming observed since 1900 and virtually a 0% of the warming observed since 1950. Is it the traditional climate model approach that is failing or is it the phenomenological model that is misleading? Is it not easy to answer this question. Either way models can be tuned to fit the data and data may be chosen to fit a pre-established opinion. However, we believe that the overwhelming evidence presented herein indicates that the phenomenological approach may be closer to the truth than the traditional climate model approach.

Chapter 5

A World of Disrupted Networks

Scientists in the eighteenth and nineteenth century recognized that the developing urban society was fundamentally different from the agrarian social order it replaced. Much of what the individual had done for himself on the farm was now done through mutual agreement in the newly interconnected social networks of the cities. The change in social order became most apparent for our purposes in the replacement of natural philosophy with quantifiable science in the nineteenth century. The origin of the word 'scientist' dates from 1833 and was coined by the philosopher and historian of science William Whewell. This term was first published in Whewell's anonymous 1834 review of Mary Somerville's On the Connection of the Physical Sciences published in the *Quarterly Review*. Curiously, at that time the term had a partly satirical meaning. In 1840 Whewell proposed the word again, this time more seriously, in his *The Philosophy of the Inductive Sciences*:

> We need very much a name to describe a cultivator of science in general. I should incline to call him a Scientist. Thus we might say, that as an Artist is a Musician, Painter, or Poet, a Scientist is a Mathematician, Physicist, or Naturalist.

A critical aspect of the new science was its reliance on experiment; a reliance that was not present in the earlier dominating perspective of Aristotle as to what constitutes knowledge and understanding. Scientists developed techniques for processing data from experiments that facilitated comparisons among investigators. The pristine methods of statistical analysis were used to smooth over the unimportant fluctuations in experimental observations and uncover patterns contained in the averages. Experimental data became the bedrock of epistemology and the quantitative nature of experiment became the true indicator of the knowable. Consequently physics entered the nineteenth century as an experimental science. However, it became apparent as that century ended that certain experimental results were inconsistent, or more accurately, the theories explaining those results were contradictory. The resolution of such contradiction revealed the importance of verifiable theory in the physical sciences, raising theory to the level of importance equal to that of experiment in the pursuit of knowledge.

In the early twentieth century the dual nature of science, theory and experiment, ushered in the new physics of quantum mechanics. The strongest evidence that the experiment-theory model of science is correct is the self-consistency of the view that theory provides the interpretation of experimental results and experiment tests theoretical predictions. The correctness of the mutually supportive aspects of theory and experiment were evident everywhere in the physical sciences and even led to a world-view in which only those theoretical concepts that could be experimentally defined were considered appropriate for scientific investigation. Even when the phenomena being investigated became more complex the challenge was met in the two-tiered context with the introduction of randomness. The observation that complex networks could not be simply predicted because they were so easily disrupted lead to the notion of multiple possible futures and probabilities.

The success of the two-tiered experiment-theory model of science was not shared by the social and life sciences as the twentieth century wore on. In fact a group of scientists argued that phenomena in physics are relatively simple and consequently the experiment-theory model is only satisfactory for their description. Not even the introduction of statistical ideas to replace the deterministic reductionism was successful is solving many of the more basic problems in the social and life sciences. It became increasingly apparent that even the simplest phenomena in the social and life sciences are truly complex and consequently the idea of proceeding directly from experimental data to theoretical knowledge was fraught with difficulty.

In the middle of the twentieth century the notion of information was introduced into this discussion using entropy as its basic metric. This definition of information was inspired since entropy provided a measure of the organization of complex physical networks. As organization within a closed network decreases the entropy increases and consequently the associated information decreases. In this way the randomness introduced to address complexity was captured by the physical quantity entropy and subsequently by information. The new concept opened the door to an extension of the scientific paradigm to one involving data, information and knowledge.

It is not a coincidence that the notion of information was developed in parallel with computer science. As analytic predictions of scientific theory were replaced by numerical predictions from complicated computer codes the information contained in the patterning of data became more accessible. The three-tiered experiment-computation-theory model of science in this way replaced the experiment-theory model of science, just as the data-knowledge paradigm was replaced with that of data-information-knowledge.

Addressing the full complexity of networks requires the analysis of information from both experiment and computation, but the transformation of data and information structures into knowledge and understanding is still the domain of theory. Leaping from data to understanding without theory often results in errors; the same can be said of extrapolating to knowledge from information and computation without theory. The present day computer codes of climate change have a degree of complexity that rivals the most complicated experiments. Consequently, the numerical output of these codes without the underlying quantitative theory has lead if not to misinterpretations, then at least to confusion.

Over the last few decades a new species of scientist has emerged; one that is self-described as being multidisciplinary. However, because schools are still tied to departments associated with disciplines and Ph.D.'s are awarded in these disciplines, the new scientists often start out as physicists, applied mathematicians, operations researchers, and so on. But they are known more by the problems they address and the techniques they use than by the titles they hold. They may be identified for purposes of academic clarity as biophysicists, econophysicists, psychophysicists, geophysicists, neuroeconomists, and so on. But even if some of the labels are familiar, the approach they take to their areas of study frequently bears little resemblance to those of their peers. For example, one of these new scientists interested in studying fields that have traditionally been the domain of biologists, sociologists, economists or geologists, will today collaborate with these field experts. In this mode of research the complexity of these phenomena are identified and the underlying network structure revealed. This transdisciplinary collaborative method of doing science has resulted in the three-tiered methodology that incorporates a careful synthesis of experimental data, computational models and theory. It is the unique capacity of this new breed of scientist to synthesize and develop fundamental principles that supersede the traditional disciplines.

This background enables the new scientists to appreciate the complex networks occurring in the heterogeneous fields of physics, biology, sociology, economics and geology and how easily they can be disrupted. They are able to see them from a perspective that is unfamiliar and often uncomfortable to the traditional practitioner. The network scientist sees these phenomena not as isolated entities separated from one another, but as realities that are connected by often subtle affinities emerging from the same fundamental laws. Laws that are not always known, and which are suggested by the common features observed in the complex networks emerging in disparate contexts.

For example, it was assumed historically that biological networks are intrinsically different from physical networks and are practically non-reducible to quantitative mathematical analysis. This view was challenged by the work of Gregor Mendel, physicist and mathematician, who proved the assumption wrong by constructing the mathematical laws that regulate genetic inheritance. When Mendel's paper was published in 1866 in the *Proceedings of the Natural History Society of Brünn*, it was ignored by the biologists of the time and was cited just three times over the next thirty-five years. Although ignored at the time, his paper is now considered a seminal work on the foundation of modern genetics. In the same spirit the nonlinear dynamical phenomenological models of today are proposed for studying and preventing the disruption of heart rhythm and explaining a variety of disruptive biomedical pathologies.

Analogously, climatologists have developed an interpretation of climate change that is essentially based on the assumption that climate cannot be significantly influenced by extraterrestrial phenomena. Up to a few years ago, TSI was considered a *constant* and this 'solar constant' was used in all computational climate models. No wonder that these climate models have found but a minor solar contribution to climate change. Alternative studies correlating climate dynamics with solar records were and are viewed with scepticism. Rightly or wrongly it appears that professional climatologists have been reluctant to incorporate astronomical mechanisms into the dynamic description of climate change, restricting analysis to a reductionistic description of purely atmospheric phenomena. Is this a masked and subtle modern form of Aristotelism: the Heavens are immutable, that is, the Sun is "constant" and the Earth is physically separated from the Heavens?

In complex networks, scientific reductionism is facilitated because of the intrinsic numerical computational complexity of the phenomena that permits models to be appropriately tuned to produce outputs that apparently agree with data. This gives the sometimes false impression that the phenomenon is understood.

For a network scientist the idea that climate may be significantly influenced by astronomical phenomena, is not a remote and unrealistic curiosity, but a possibility that should be carefully investigated. Empirical evidence of solar variability and its disruptive influence on climate have become increasingly firm over the last two decades. The fact that the exact Sun-climate linking mechanisms are still not fully understood does not imply that those mechanisms do not exist. After all, it is certainly plausible that astronomical perturbations may effect climate change. Moreover nonlinear dynamics have shown us the emergence of many surprising phenomena, and, as Newton established, the Earth and the Heavens may have much more in common than what an apparent *consensus* among any group may lead people to believe.

In this book we have briefly discussed how a three-tiered science naturally emerges when complex networks are analyzed.

In Chapter 2 the fundamental idea of data was explored from a historical perspective. Experiments provide data about individuals in psychophysics, about groups in sociophysics, about the exchange of commodities in econophysics and about life in biophysics. In this way the dynamics from a variety of complex adaptive networks are manifest in time series data. We used a number of physiological time series as exemplars of phenomena that have the history of the organ built into their time variation.

The importance of time series data is nowhere more evident than in the global warming debate. We used simple statistical measures to show the interdependence of the average global temperature time series and the solar activity as measured by sunspot and solar flare time series. But we also discussed some of the difficulties associated with obtaining reliable data sets on a global scale, as well as the problems with using such time series to draw conclusions. A quite remarkable comparison was that between the statistics of the average global temperature anomalies and the fluctuations in the total solar irradiance. We showed that the Earth's average temperature fluctuations inherit the exotic statistics of the fluctuations in solar activity.

In Chapter 3 various measures of complex dynamical networks characterized by the time series from the previous chapter were discussed, as well as how these measures influence the design and understanding of networks.

Our emphasis was on entropy as a measure of information; the patterning or organization found in experimental data. The information generated by the nonlinear dynamics of a complex adaptive network should be independent of the specific mechanism doing the generating and therefore the information may provide insight into possible universal properties of complexity. One such universal property relates to statistics. The Normal or Gaussian distribution was shown to be too restrictive to describe the range of behavior accessible to complex adaptive networks and is replaced by the inverse power-law distribution of Pareto. Some example applications of the Pareto distribution to economic, informational, neuronal, science, social and terrorist networks were discussed. In particular, the greater probability of events far from the average in the social and life sciences implies a much greater dependence of these complex phenomena on potentially disruptive events than is observed in simpler physical phenomena.

The scaling of the probability distribution function encapsulates the structure of the network complexity. Of special interest was how one complex network perturbs the dynamics of a second complex network, resulting in the network of interest inheriting the statistics of the perturbing network. This general theoretical notion found application in the variability of the Sun's solar activity perturbing the fluctuations in the Earth's average global temperature. In Chapters 2 and 3 both climate and solar proxy reconstruction data were shown to be uncertain, but the information inferred from them is that there exists a linking between solar dynamics and global climate and this linking can be quite significant.

Finally, in Chapter 4 we noted that since data is the outcome of experiment and information is the order contained in the data that knowledge is the interpretation of the order in the data using theory. In Chapter 4 we attempted to go beyond theory in the context of a given discipline and associated theory with the complex adaptive network underlying the phenomenon of interest. The existence of such a discipline-independent theory would imply that a Network Science can also exist. A Network Science would supersede the traditional scientific disciplines, while being reducible to any of them in an appropriately restricted situation. However, we determined that the inverse power-law distribution does not have a unique mathematical generator, but is entailed by a wide variety of mechanisms.

If the network structure represents the most appropriate phenomenological description of a complex system, there does not exist a unique model for it. Topologies alone are not sufficient to describe how a network behaves.

Nodes and links both play their roles. In fact, although there may be an average six-degrees of separation between two generic people in the world, it is not true that real people are so simply connected. The network is indeed disrupted because what really determines the behavior of a network is how information flows, not just the network's *potential* topology. If no information is exchanged between two nodes, no real link exists between them: it may be true that an average American citizen is separated from Osama Bin Laden by just six contacts, but that is not sufficient to make him/her a potential terrorist, collaborator, or even somebody that knows anything about Osama Bin Laden!

How information is generated, transmitted and processed by and within a web incorporating its subnetworks is the true scientific challenge of the twenty-first century. A challenge that, if overcome, will likely yield a great unification of all fields of scientific knowledge.

There exists another surprising reason supporting the observation that we are living in a world of *disrupted networks*. The remarkable example is that a network of knowledge can be disrupted by the simple linking of that network to another network; one that is a possibly inhomogeneous network of knowledge. But how are we to interpret this sensitivity?

Today we live in a world of increasingly open access to information. Since ancient times the transmission of knowledge has been considered a major factor in the progress of the knowledge itself. In this prehistory the major repository of knowledge were the elders; the old, mostly men, that instructed the young teaching them society's secrets. This custom fostered a long oral tradition whereby the knowledge of the fathers continuously accreted to the sons in the next generation. Evidently the growth rate of knowledge transmitted in this way was limited, even without the filtering of what was passed along by personal choices and capacities. Consequently human civilizations as measured by the growth and accumulation of knowledge progressed slowly.

Eventually people developed a more efficient tradition and collected past knowledge into books and collected these books into libraries. This 'information technology' allowed new generations and an increasing number of people access to higher specialized knowledge without the distorting filter of multiple sequential memories. The revolution of written traditions spawned what is properly called *human history*. The clusters of human civilizations that adopted this way of storing knowledge became linked to one another and progressed faster than those that had no such repositories. Written traditions gave rise to the first knowledge revolution in human history, disrupting the family group or clan that had persisted for untold generations through their oral networks.

The 20th century has greatly expanded the knowledge network by means of media such as radio, television and a wide and relatively cheap production of books and magazines. Several hundred million people had access to a vast storehouse of knowledge that never existed previously. This larger network has contributed to the rapid and significant progress humanity has experienced during the last century.

The 21st century will be characterized by an even larger network: the World Wide Web and the Internet. Today an incredibly large amount of information already freely flows on the web and several hundred million people world wide can almost instantaneously and with relatively modest economic means access these resources. These individuals can and do contribute to the development of new knowledge by applying their expertise to the data and information already available on the Internet.

Such a network of knowledge may be disrupted any time a new link or node is added to it. Links and nodes are in this case people that contribute to the general culture with new ideas. In science it is common place for a single new researcher to contribute to a discipline. Once in a great while that contribution is revolutionary; turning accepted knowledge on its head. Scientists such as Galileo, Newton and Einstein were such rare people. Einstein was even an outsider of the well established academia network of physicists. He was an obscure employee at a Swiss patent office that could not find a teaching position at any university. It was as a patent clerk that Einstein disrupted the physical notions of space and time with his special theory of relativity. It is not necessary to mention the other astonishing discoveries he made in statistical physics and quantum theory at essentially the same time and would also subsequently transform their respective areas of physics.

The World Wide Web has the potential for disrupting established knowledge networks simply because outsiders and non-specialists have access to it and can exchange and enhance knowledge, without being filtered by the elders. It is not a coincidence that authoritarian governments have always attempted to control and if not control then limit the circulation and exchange of ideas. Today such governments are particularly intimidated by the Internet and therefore block access to web-site that are judged dangerous to the establishment.

In a related way an academy may, with the best of intentions, become closed to new ideas rendering it scientifically sterile by favoring one point of view over another about a given phenomenon and prejudging the results of analysis. One way this rigidity is manifest is by selecting new faculty on the basis of their conformity to a given scientific view. Those who disagree and propose alternative interpretations are shunned or even worse simply ignored.

This process of sterilization had been occurring at a number of institutions with the closing of debate over climate change. Fortunately, however the fading debate over climate change was disrupted in part, because it attracted the attention of amateurs on the World Wide Web. Numerous web blogs are written to extensively discuss peer reviewed literature on this topic by talented amateurs and professional scientists alike. These web-blogs (for example: http://wattsupwiththat.com/, http://rankexploits.com/musings/, http:// www.climateaudit.org/ and http://www.realclimate.org/) with their own limits have the merit of making this intriguing issue available to all. Many people, both inside and outside of academia and the world of professional research, contribute to these web-blogs with their comments, their ideas and sometime even with their our research. It is not possible to rule out that valid ideas or even a resolution to the climate change debate may be proposed by such amateurs that study this issue as a hobby.

The advent of network complexity has revealed the intrinsic difficulty in developing an unified theory capable of interpreting the whole of reality. The networks of data, the webs of information and latticework of knowledge are being continuously disrupted by virgin data, fresh information and new knowledge. However, this does not diminish our hope for finding a richer understanding of reality, one that encompasses the dynamics of change while leaving undiminished our awe of Nature.

Glossary

Albedo: fraction of incident solar radiation that is reflected directly back to space. Earth's albedo equals about 30%.

Astronomical unit: the average distance between the Earth and the Sun, about 150 million kilometers, or 93 million-miles.

Atmosphere of the Earth: is made of 77% nitrogen, 21% oxygen, 1% argon, carbon dioxide, water vapor and other trace gases.

Attractor: a geometrical structure in phase space to which all the trajectories describing the evolution of a network are drawn: Hence the name attractor.

Average: mathematically it is the sum over all the values in a data set divided by the total number of data points; it is presumed to typify the data.

Bit: a contraction of binary digit used to mean a unit of information coded as a physical state of a two-state system.

Butterfly effect: denotes the sensitivity of a network's dynamics to its initial state, such as the flapping of a butterfly's wings in New York causing a hurricane in Brazil.

Carbon dioxide (CO_2): is formed by combustion of fossil fuels, decay of biological matter and by breathing of animals and plants. It is classified as a greenhouse gas.

Central limit theorem: the mathematics establishing the conditions under which the statistics of a complex network are described by a single humped normal distribution with finite mean and variance.

Chaos: the extremely sensitive dependence of a network's dynamics on the initial state of the network, thereby limiting the time over which the network's evolution can be predicted.

Complexity: much of the discussion in this book is devoted to a definition of this term; but it typically involves many elements, interacting in unpredictable ways.

Correlation function: a measure of the influence a fluctuation in a time series has on other fluctuations in a time series. The correlation function describes the mathematical form of that influence as a function of the time separation between fluctuations.

Cosmic rays: energetic ionized nuclei and electrons in space and are both galactic and solar in origin.

Cybernetics: is the science of how humans exchange information with machines.

Dalton minimum: prolonged sunspot minimum from 1795 to 1823.

Data: the recorded measurements resulting from experiment and/or observation.

Disorder: the degree of irregularity or unpredictability in the structure of a network.

Dissipative structure: a pattern within a dynamical system that is maintained by the flux of energy or matter through the system; the entropy of such a structure decreases relative to the environment over time.

Dynamical networks: change over time. Usually this name is reserved for networks whose time changes are described by deterministic dynamical equations, but the equations of motion might be statistical as well.

Econophysics: the systematic application of the quantitative methods of the physical sciences to economic networks.

Energy balance models: simpler climate models that balance, or very nearly balance, incoming energy as short wavelength electromagnetic radiation to the Earth with outgoing energy as long wavelength electromagnetic radiation from the Earth. Any imbalance results in a change in the average temperature of the Earth.

Entropy: a state-dependent network function that measures the amount of energy in a system that is not available to do work; alternatively a measure of the amount of disorder (information) in a network.

Environment: the context in which one finds a network.

Ensemble: an infinitely large number of copies of the orbits describing a network's dynamics used to construct a probability density.

Ergodic theory: a branch of mathematics relating probability measures to dynamical networks. A network is ergodic when a long-time average over a single orbit is the same as the average over an ensemble of orbits at a single time.

Extensivity: the additive nature of network properties with the size of the network. For example, energy is extensive since doubling the size of a homogeneous network doubles the energy.

Flares: brief increase in brightness of a small solar region. Flares yield intense bursts of ultraviolet radiation and, sometime can be observed in white light.

Fractal: an object whose parts resemble the whole in some way.

Fractal dimension: the non-integer dimension of a fractal object or process.

Fractal geometry: the study of the geometrical properties of fractal objects.

Fractal statistics: the study of the statistical properties of fractal time series.

General circulation model: model of the Earth's atmosphere first developed for weather predictions and later extended to predict climate change.

Graph theory: the study of the properties of network using the connectivity between elements.

Greenhouse effect: the blanketing effect of some gases which are opaque to thermal radiation, and whose accumulation warms the Earth. Major greenhouse gases are water vapor, carbon dioxide, methane, ozone, freon.

Hale cycle: the variation of the sun's magnetic field over a period of approximately 22 years, during which the field reverses and is restored to its original polarity; one such cycle comprises two successive sunspot cycles.

Ice ages: periods of extreme cold when glaciers advanced and covered much of the Earth's land area. There were about 20 ice ages during the last 2 million year. They may be explained by the Milankovitch Hypothesis.

Information: the patterns observed in data sets.

IPCC: the Intergovernmental Panel on Climate Change (IPCC) was established by the United Nations Environmental Programme (UNEP) and the World Meteorological Organization (WMO) in 1988 to assess the scientific, technical and socio-economic information relevant for the understanding of human induced climate change, its potential impacts and options for mitigation and adaptation. The IPCC has completed four full assessment reports, guidelines and methodologies, special reports and technical papers.

Knowledge: the interpretation of patterns in data sets using theory.

Lévy distribution: when the variance of a data set diverges, an extension of the central limit theorem yields a class of probability densities with diverging variance.

Map: is a mathematical procedure for taking a set of numbers and systematically transforming them into a different set of numbers, say from one point in time to another point in time.

Maunder minimum: prolonged sunspot minimum from 1645 to 1715.

Memory: is used in the technical sense of correlated fluctuations, usually associated with a long-time, inverse power-laws.

Milankovitch Hypothesis: a theory for explaining the timing of the ice ages. It suggests that the Earth's climate varies according to the variation if its orbital characterisitics such as precession, obliquity and eccentricity.

Nonlinear dynamics: the mathematical equations describing the time evolution of a network whose elements do not respond in proportion to their interaction with other elements in the network.

Negentropy: a term coined by Schrödinger to denote the entropy an ordered network extracts from the environment necessary to retain its order.

Network: consists of a set of elements (variables) that interact with each other through a set of relations among the elements in a well-defined manner.

Noise: a data set consisting of unique elements that are independent of one another.

Objective: The presumption that what we know and understand about a network can be separated from the person that is observing the network. Although knowledge is determined by the interaction of the observer with the observed, there does exist a world that is independent of the observer and is therefore independently knowable.

Orbit: the trail of states occupied by a network in phase space as the dynamics of the network unfolds.

Order: the degree of regularity in a network, often determined by the degree of predictability in a network's dynamics.

Paleoclimatic temperature reconstruction: reconstruction of the global surface temperature of the Earth by using data from ice cores, tree rings, sediments, pollen and whatever else is available.

Paradox: the apparent contradiction of two results.

Phase space: the space whose coordinates constitute the independent variables required to fully describe the properties of a network. A point in this space gives a complete specification of the state of the network.

Probability: the relative number of times an event can occur relative to all the possible outcomes of a process.

Probability density: one cannot specify the probability of a particular event for a continuous process; rather, one specifies the probability of events occurring within an interval. The probability of an outcome occurring within that interval is the product of the probability density and the width of the interval.

Psychophysics: the systematic application of the quantitative methods of the physical sciences to psychological networks.

Random: A process or event is completely random when it cannot be predicted from the previous history of the process, or from the conditions of the environment just prior to the event.

Random walk: steps taken on a lattice where the direction of successive steps are random.

Reductionism: is entailed when the properties of a network are implied by the fundamental laws governing the basic constituents, or elements, of the network.

Schwabe cycle: 11-year solar cycle observed in the sunspot number record.

Small world theory: describes the empirical observation that individuals in a very large social network are connected by very many fewer interactions than had been previously thought, for example, by 'six degrees of separation'.

Solar activity: refers to the whole of the solar phenomena such as sunspot number, faculae, total solar irradiance, magnetic activity etc.

Spoerer minimum: prolonged sunspot minimum from 1450 to 1540.

Sociophysics: is the systematic application of the quantitative methods of the physical sciences to sociology networks.

Stratosphere: the region of the Earth's atmosphere above the troposphere.

Sunspots: darker region in the sun's photosphere. These regions vary in number and area with an 11-year solar cycle.

Strange attractor: a structure in phase space to which trajectories are attracted and on which the dynamics of a chaotic network unfolds asymptotically. The dimension of such an attractor is not integer.

Subjective: the presumption that what we know and understand about a network cannot be separated from the person that is observing the network. All knowledge is determined by the interaction of the observer with the observed and is therefore not independently knowable.

Time series: a set of random data, with each data point being measured at a successive time point.

Total solar irradiance: the sun's total omnidirectional energy outflows. It is known also as solar luminosity. It is about $1366\ W/m^2$ and varies of about $1\text{-}2\ W/m^2$. It is measured with detectors on board of satellites since 1978.

Trajectory: the same as a phase space orbit.

Troposphere: the lower region of the Earth's atmosphere above the surface with the Hadley cell circulation and vertical convection. It extends 15 km (about 10 miles) on the equator and about 8-10 km at the poles.

Variance: a measure of the variability of a data set; related to the width of the probability density.

Bibliography

[1] Abbott D., (1984), *The Biographical Dictionary of Scientists: Physicists*, Peter Bedrick Books, New York.

[2] Amaral L. A. N., A. Scala, M. Barthélémy and H. E. Stanley, (2000), "Classes of small-world networks," *Proc. Natl Acad. Sci. USA* **97**, 11149-11152.

[3] Albert R., H. Jeong and A. Barabasi, (2000), "Error and attack tolerance of complex networks," *Nature* **406**, 378-382.

[4] Alderson D., L. Li, W. Willenger and J. C. Doyle, (2005), "Understanding Internet Topology: Principles, Models and Validation," *IEEE/ACM Transactions on Networking* **13**, 1205-1218.

[5] Allegrini P., M. Bologna, P. Grigolini and B.J. West, (2007), "Fluctuation-dissipation theorem for event-dominated processes," *Physical Review Letter* **99**, 010603.

[6] Arrhenius S. A., (1896), "On the Influence of Carbonic Acid in the Air upon the Temperature of the Ground," *Philosophical Magazine* **41**, 237-276.

[7] Ball D. A., (1956), *Information Theory*, 2nd Edition, Pitman, New York.

[8] Barabási A.-L., (2003), *Linked*, Plume, New York.

[9] Albert R. and Barabási A.-L., "Statistical mechanics of complex networks," *Rev. Mod. Phys.* **74**, 48 (2002).

[10] Bennett C. H., (1982), "The thermodynamics of computation — a review," *International Journal of Theoretical Physics* **21**, 905-940;

[11] Bennett C. H., (1987), "Demons, engines and the 2nd law," *Scientific American*, v. 257, p. 108.

[12] Berry M. V., (1981), "Quantizing a classically ergodic system — sinai billiard and the KKR method," *Annals of Physics* **131**, 163-216.

[13] von Bertalanffy L., (1968), *General Systems Theory*, G. Braziller, New York.

[14] Brillouin L., (1962), *Science and Information Theory*, Academic Press, New York.

[15] Bochner S., (1966), *The Role of Mathematics in the Rise of Science*, Princeton University Press, Princeton, NJ.

[16] Boffetta G., V. Carbone, P. Giuliani, P. Veltri and A. Vulpiani, (1999), "Power laws in solar flares: Self-organized criticality or turbulence?," *Physical Review Letter* **83**, 4662-4665.

[17] Brohan P., J. J. Kennedy, I. Haris, S. F. B. Tett and P. D. Jones, (2006), "Uncertainty estimates in regional and global observed temperature changes: a new dataset from 1850," *J. Geophysical Research* **111**, D12106. DOI:10.1029/2005JD006548.

[18] Brooks C. E. P., (1970), *Climate through the Ages*, 2nd revised edition, Dover, New York; original publication data 1926 and revised in 1940.

[19] Buchanan M., (2002), *Nexus*, Norton & Comp., New York.

[20] Bush V., (1945), Science The Endless Frontier, A Report to the President by Vannevar Bush, Director of the Office of Scientific Research and Development, July 1945: United States Government Printing Office, Washington. *http://www.nsf.gov/od/lpa/nsf50/vbush1945.htm*

[21] Camp C. D., and K. K. Tung, (2007), "Surface warming by the solar cycle as revealed by the composite mean difference projection," *Geophysical Research Letterts* **34**, L14703, doi:10.1029/2007GL030207.

[22] Carlson J. M. and J. C. Doyle, (1999), "Highly Optimized Tolerance: a mechanism for power laws in designed systems," *Physical Review E* **60**, 1412-1427.

[23] Casti J. L., (2002), "Biologizing Control Theory: How to make a control system come alive," *Complexity* **7**, 10-12.

[24] Clausius, R. (1850), "Über die bewegende Kraft der Wärme, Part I, Part II," *Annalen der Physik* **79**: 368-397, 500-524 . See English Translation: "On the Moving Force of Heat, and the Laws regarding the Nature of Heat itself which are deducible therefrom." *Phil. Mag.* (1851), **2**, 121, 102-119.

[25] Clayton H. H. (1923), *World Weather, including a discussion of the influence of variations of the solar radiation on weather and of the meteorology of the Sun,* Macmillan, New York.

[26] Cohen I. B., (1987), "Scientific Revolutions, Revolutions in Science, and a Probabilitistic Revolution 1800-1930," in *The Probabilistic Revolution: Vol.1 Ideas in History,* MIT Press, Cambridge MA.

[27] Committee on Department of Defense Basic Research, National Research Council (2005), *Assessment of Department of Defense Basic Research,* The National Academy of Sciences.
http://www.nap.edu/openbook/0309094437/html/1.html.copyright

[28] Committee on Network Science for Future Army Applications, National Research Council (2005), *Network Science,* The National Academy of Sciences.

[29] Cook E. R., D. M. Meko and C. W. Stockton, (1997), "A new assessment of possible solar and lunar forcing of the bidecadal drought rhythm in the Westrern United States," *Journal of Climate* **10**, 1343-1356.

[30] Cox P. M., Betts R. A., Jones C. D., Spall S. A. and Totterdell I. J., (2000), "Acceleration of global warming due to carbon-cycle feedbacks in a coupled climate model," *Nature* **408**, 184-187.

[31] Coughlin K. and K. K. Tung, (2004), "11-year solar cycle in the stratosphere extracted by the empirical mode decomposition method," *Solar Variability and Climate Change* **34**, 323-329.

[32] Coughlin K. and K. K. Tung, (2004), "Eleven-year solar cycle signal throughout the lower atmosphere," *Journal of Geophysical Research* **109**, D21105.

[33] Cowan G. A., D. Pines and D. Meltzer, (1994), *Complexity: Metaphors, Models and Reality*, Santa Fe Institute Studies in the Science of Complexity, Addison-Wesley, Reading, MA.

[34] Crooks S. A. and L. J. Gray, (2005), "Characterization of the 11-year solar signal using a multiple regression analysis of the ERA-40 dataset," *Journal of Climate* **18**, 996-1015.

[35] Crowley T. J. and K. Y. Kim (1996), "Comparison of proxy records of climate change and solar forcings," *Geophysical Research Letters* **23**, 359-362.

[36] Crowley T. J., (2000), "Causes of Climate Change Over the Past 1000 Years," *Science* **289**, 270-277.

[37] Climatic Research Unit, UK. *http://www.cru.uea.ac.uk.*

[38] Davis H. T., (1941), *The Theory of Econometrics* (Bloomington Ind.)

[39] Devlin K., (2002), *The Millennium Problems*, Basic Books, NY.

[40] Donarummo J. Jr., M. Ram, and M. R. Stolz, (2002), "Sun/dust correlations and volcanic interference," *Geophysical Research Letters* **29**, 1361-1364, doi:10.1029/2002GL014858.

[41] Douglass D. H., and B. D. Clader, (2002), "Climate sensitivity of the Earth to solar irradiance," *Geophysical Research Letters* **29**, 1786-1789, doi:10.1029/ 2002GL015345.

[42] Dzurisin D., (1980), "Influence of Fortnightly Earth Tides at Kilauea Volcano, Hawaii," *Geophysical Research Letters* **7**, 925-928.

[43] Ebeling W. and G. Nicolis, (1991), "Entropy of symbolic sequences — the role of correlations," *Europhysics Letters* **14**, 191-196.

[44] Ebeling W., (1993), "Entropy and information in processes of self-organization — uncertainty and predictability," *Physica A* **194**, 563-575.

[45] Eddy J. A., (1976), "The Maunder minimum," *Science* **192**, 1189-1202.

[46] Erdős, P. and A. Rényi, "Random Graphs" in *Mathematical Institute of the Hungarian Academy of Science* **5**, 17-61 (1960).

[47] Eve R. A., S. Horsfall, and M. E. Lee (eds), (1997), *Chaos, Complexity and Sociology*, SAGE , Thousand Oaks.

[48] Faloutsos M., P. Faloutsos and C. Faloutsos, (1999), "On Power-law Relationship of the Internet Topology," *Computer Communication Review* **29**, 251.

[49] Fechner G. T., (1860), *Elemente der Psychophysik*, Breitkopf und Hartel, Leipzig.

[50] Fix J. D., (2001), *Astronomy, Journey to the Cosmic Frontier*, 2nd edition, McGraw-Hill, New Jork.

[51] Foukal P., G. North and T. Wigley, (2004), "A Stellar View on Solar Variations and Climate," *Science* **306**, 68-69.

[52] Fröhlich C. and J. Lean, (1998), "The Sun's total irradiance: Cycles, trends and related climate change uncertainties since 1976," *Geophysical Research Letters* **25**, 4377-4380.

[53] Frieman E.A., (1994), "Solar variation and climate change" in *Solar Influence on Global Change*, National Academy Press, Washington, D.C. 23-47.

[54] Friis-Christensen E. and K. Lassen, (1991), "Length of the solar cycle: an indicator of solar activity closely associated with climate," *Science* **254**, 698-700.

[55] Friis-Christensen E., C. Fröhlich, J.D. Haigh, M. Schüssler, and R. von Steiger, (2000), *Solar Variability and Climate*, Kluwer Academic Publisher, London.

[56] Gauss F., (1809), *Theoria motus corporum coelestrium*, Hamburg, Dover Eng. Trans.

[57] Giuliani P., V. Carbone, P. Veltri, G. Boffetta and A. Vulpiani, (2000), "Waiting time statistics in solar flares," *Physica A* **280**, 75-80.

[58] Gleisner H. and P. Thejll, (2003), "Patterns of tropospheric response to solar variability," *Geophysical Research Letters* **30**, 1711.

[59] Goldberger A. L., (1996), "Non-linear dynamics for clinicians: chaos theory, fractals, and complexity at the bedside," *Lancet* **347**, 1312-1314.

[60] Graunt J. (1936), *Natural and Political Observations Mentioned in a Following Index and Made Upon the Bills of Mortality*, London (1667), Reprint Edition, Johns Hopkins.

[61] Grigolini P., D. Leddon and N. Scafetta, (2002), "The Diffusion entropy and waiting time statistics of hard x-ray solar flares," *Physical Review E* **65**, 046203.

[62] Hansen J. et al., (2007), "Climate simulations for 1880-2003 with GISS modelE," *Climate Dynam.* **29**, 661-696, doi:10.1007/s00382-007-0255-8.

[63] Ho L., (August 2004), lecture at ARO symposium on "Complex Adaptive Systems".

[64] Hodell D. A., M. Brenner, J. H. Curtis, and T. Guilderson, (2001), "Solar Forcing of Drought Frequency in the Maya Lowlands," *Science* **292**, 1367-1370.

[65] Houghton J. T., (2001), *et al.*, *Intergovernmental Panel on Climate Change, Climate Change 2001: The Scientific Basis*, Cambridge University Press, UK. *http://www.ipcc.ch*.

[66] Hoyt D. V. and K. H. Schatten, (1997), *The role of the Sun in the Climate Change*, Oxford University Press, New York.

[67] Hoyt D.V. and K. H. Schatten, (1998), "Group Sunspot Numbers: A new solar activity reconstruction," *Solar Physics* **181**, 491-512.

[68] Hurst H., (1951), "Long Term Storage Capacity of Reservoirs," *Transactions of the American Society of Civil Engineers* **116**, 770-799.

[69] Hurst H. E., R. P. Black, Y. M. Simaika, (1965), *LongTerm Storage: An Experimental Study*, Constable, London.

[70] Juran J., (1988), *Quality Control Handbook*, 5th Edition, McGraw-Hill, New York, New York.

[71] Kelso S., (August 2004), lecture at ARO symposium on "Complex Adaptive Systems".

[72] Khinchine A. I., (1957), *Mathematical Foundations of Information Theory*, trans. R. A. Silverman & M. D. Friedman, Dover, New York.

[73] Kirkby J., (2007), "Cosmic Rays and Climate," *Surveys in Geophysics* **28**, 333-375.

[74] Kleinfeld J. S. *Could it be a big world after all? The 'six degrees of separation' myth*, http://www.uaf.edu/northern/big_world.html.

[75] Koch R., (1998), *The 80/20 Principle: The Secret of Achieving More with Less*, Nicholas Brealey Publishing.

[76] Krebs V., (2008), "*Connecting the Dots Tracking Two Identified Terrorists*," http://www.orgnet.com/tnet.html.

[77] Krivova N. A., L. Balmaceda, S. K. Solanki, (2007), "Reconstruction of solar total irradiance since 1700 from the surface magnetic flux," *Astronomy and Astrophysics* **467**, 335-346.

[78] Kubo R. (1957), "Statistical-Mechanical Theory of Irreversible Processes. I. General Theory and Simple Applications to Magnetic and Conduction Problems," *Journal Physical Society of Japen* **12**, 570-586.

[79] Kubo R., M. Toda and N. Hashitusume (1985), *Statistical Physics*, Springer, Berlin.

[80] Lamb H. H., (1995), *Climate, History and the Modern World*, Methuen, London.

[81] Landauer R., (1961), "Irreversibility and heat generation in the computing process," *IBM J. Res. Develop.* **5**, 183-191.

[82] Lasota A. and M. C. Mackey, (1994), *Chaos, Fractals and Noise*, Springer-Verlag, New York.

[83] Lean J., J. Beer, and R. Bradley, (1995), "Reconstruction of solar irradiance since 1610: implications for climate change," *Geophysical Research Letters* **22**, 3195-3198.

[84] Lean J., (2000), "Evolution of the Sun's spectral irradiance since the Maunder Minimum," *Geophysical Research Letters* **27**, 2425-2428.

[85] Lean J. and D. Rind, (2001), "Earth's Response to a Variable Sun," *Science* **292**, 234-236.

[86] Lean J., (2005), "Living with a variable Sun," *Physics Today* **58**, 32-38.

[87] Lewin B. (ed.), (1998), "Sex i Sverige. Om Sexuallivet i Sverige 1996 [Sex in Sweden. On the Sexual Life in Sweden 1996]," *Natl Inst. Pub. Health*, Stockholm.

[88] Liljeros F., C. R. Edling, L. A. N. Amaral, H. E. Stanley, and Y. Aberg, (2001), "The web of human sexual contacts," *Nature* **411**, 907-908.

[89] Lindenberg K.L. and B.J. West, (1990), *The Nonequilibrium Statistical Mechanics of Open and Closed Systems*, VCH, New York, New York.

[90] Lisiecki L. E., and M. E. Raymo (2005), "A Pliocene-Pleistocene stack of 57 globally distributed benthic $\delta^{18}O$ records," *Paleoceanography* **20**, PA1003, doi:10.1029/2004PA001071.

[91] Lockwood M. and R. Stamper (1999), "Long-term drift of the coronal source magnetic flux and the total solar irradiance," *Geophys. Res. Lett.* **26**, 2461-2464.

[92] van Loon H. and D. J. Shea, (1999), "A probable signal of the 11-year solar cycle in the troposphere of the northern hemisphere," *Geophysical Research Letters* **26**, 2893-2896.

[93] van Loon H. and K. Labitzke, (1999), "The signal of the 11-year solar cycle in the global stratosphere," *Journal of Atmospheric and Solar-Terrestrial Physics* **61**, 53-61.

[94] van Loon H. and D. J. Shea, (2000), "The global 11-year solar signal in July-August," *Geophysical Research Letters* **27**, 2965-2968.

[95] van Loon H. and K. Labitzke (2000), "The influence of the 11-year solar cycle on the stratosphere below 30 km: A review," *Space Science Review* **94**, 259-278.

[96] Lorenz E. N., (1993), *The Essence of chaos*, University of Washington Press.

[97] Lotka A., (1924), *Elements of Physical Biology*, Williams and Wilkins Co., New York.

[98] Lotka A. J., (1926), "The Frequency Distribution of Scientific Productivity," *J. Wash. Acad. Sci.* **16**, 317-320.

[99] MacDonald N., (1983), *Trees and Networks in Biological Models*, John Wiley & Sons, New York.

[100] Mackey M. C., (1992), *Time's Arrow*, Springer Verlag, New York.

[101] May R. M., (1976), "Simple mathematical models with very complicated dynamics," *Nature* **261**, 459-467.

[102] Mandelbrot B. B., (1977), *Fractals, form, chance and dimension*, W.H. Freeman and Co., San Francisco.

[103] Mandelbrot B. B., (1982), *The Fractal Geometry of Nature*, W. H. Freeman and Co., San Francisco.

[104] Mann M. E., R. S. Bradley, and M. K. Hughes, (1999), "Northern Hemisphere temperatures during the past millennium: inferences, uncertainties, and limitations," *Geophysical Research Letters* **26**, 759-762.

[105] Mann M. E. and P. D. Jones, (2003), "Global Surface Temperatures over the Past Two Millennia," *Geophysical Research Letters* **30**, 1820-1823. doi:10.1029/2003GL017814

[106] Marinov I., Gnanadesikan A., Toggweiler J. R. and Sarmiento J. L., (2006), "The Southern Ocean biogeochemical divide," *Nature* **441**, 964-967.

[107] McKitrick R., and P. Michaels, (2007), "Quantifying the influence of anthropogenic surface processes and inhomogeneities on gridded global climate data," *J. Geophys. Res.* **112**, D24S09, doi:10.1029/2007JD008465.

[108] Meakin P., (1998), *Fractals, scaling and growth far from equilibrium*, Cambridge Nonlinear Science Series 5, Cambridge University Press, New York.

[109] Meldrum C., (1872), "On a periodicity in the frequency of cyclones in the Indian Ocean south of the equator," *Nature* **6**, 357.

[110] Milankovitch M., (1920), *Theorie Mathematique des Phenomenes Thermiques produits par la Radiation Solaire.* Gauthier-Villars Paris.

[111] Mill J. S., (1843), *System of Logic, Ratiocinative and Inductive*, reprinted edition, University Press of the Pacific, Honolulu, 2002.

[112] Mitchell J. M., C. W. Stockton, and D. M. Meko (1979), "Evidence of a 22-year rhythm of drought in the western United States related to the Hale solar cycle since the 17th century," *Solar-Terrestrial Influences on Weather and Climate*, B.M. McCormac and T.A. Seliga, Eds., D. Reidal, p. 125-144.

[113] Moberg A., D. M. Sonechkin, K. Holmgren, N. M. Datsenko and W. Karlén, (2005), "Highly variable Northern Hemisphere temperatures reconstructed from low- and high-resolution proxy data," *Nature* **443**, 613-617. doi:10.1038/nature03265

[114] de Moivre A., (1967), *Doctrine of Chances; or A Method of Calculating the Probabilities of Events in Play*, London, (1718); 3rd edition (1756); reprinted by Chelsea Press.

[115] Montroll E. W. and W. W. Badger, (1974), *Introduction to the Quantitative Aspects of Social Phenomena*, Gordon and Breach, New York.

[116] Murray C.D., (1927), "A relationship between circumference and weight and its bearing on branching angles," *Journal General Physiology* **10**, 125-729.

[117] Muscheler R. *et al.*, (2007), "Solar activity during the last 1000 yr inferred from radionuclide records," *Quaternary Science Reviews* **26**, 82-97.

[118] Neuberger H., (1970), "Climate in Art," *Weather* **25**, 46-56.

[119] Nicolis J. S., (1991), *Chaos and Information Processing*, World Scientific, Singapore.

[120] North G., *et al.*, (2006), "Surface Temperature Reconstructions for the Last 2,000 Years," *The National Academies Press*.

[121] Oberschall A., (1987), "Empirical Roots of Social Theory," in *The Probabilistic Revolution, Volume 2, Ideas in the Sciences*, Eds. Krüger L., G. Gigerenzer and M.S. Morgan, MIT Press, Cambridge.

[122] Onsager L., (1931), "Reciprocal Relations in Irreversible Processes. I," *Physics Review* **37**, 405-426; "Reciprocal Relations in Irreversible Processes. II" *Physics Review* **38**, 2265-2279.

[123] Pap J. M., *et al.*, (2004) (EDS), *Solar Variability and its Effects on Climate*, Geophysical Monograph Series Volume 141 American Geophysical Union, Washington, DC.

[124] Pareto V., (1896), *Cours d'e'conomie politique*, Laussane.

[125] Penrose R., (1994), *Shadows of the Mind, a Search for the Missing Science of Consciousness*, Oxford University Press, Oxford.

[126] Petit J. R. *et al.*, (1999), "Climate and Atmospheric History of the Past 420,000 years from the Vostok Ice Core, Antarctica," *Nature* **399**, 429-436.

[127] Poincaré H., (1913), *The Foundations of Science*, The Science Press, New York.

[128] Quetelet A., (1848), *Du Système Social et des Lois Qui le Règissent*, Paris.

[129] Quinn T. R. *et al.*, (1991), "A Three Million Year Integration of the Earth's Orbit," *The Astronomical Journal* **101**, 2287-2305.

[130] Rall W., (1959), "Theory of physiological properties of dendrites," *Annals of NewYork Academy of Science* **96**, 1071-1091.

[131] Redner S., (1998), "How popular in your paper? An empirical study of the citation distribution," *Eur. Phys. Jour. B* **4**, 131-134.

[132] Reid G. C., (1991), "Solar total irradiance variations and the global sea-surface temperature record," *Journal of Geophysical Research* **96**, 2835-2844.

[133] RGO (2003), Royal Greenwich Observatory/USAF/NOAA Sunspot Record 1874-2003.
http://science.msfc.nasa.gov/ssl/pad/solar/greenwch.htm

[134] Richter J. P., Editor, *The Notebooks of Leonardo da Vinci, Vol.1*, Dover, New York; unabridged edition of the work first published in London in 1883.

[135] Rind D. D., (2002), "The Sun's role in climate variations," *Science* **296**, 673-677.

[136] Roberts F. S., (1979), *Measurement Theory, Encyclopedia of Mathematics and Its Applications*, Addison-Wesley, London.

[137] Ruddiman W. F., (2003), "The anthropogenic greenhouse era began thousands of years ago," *Climatic Change* **61**, 261-293.

[138] Scafetta N., P. Hamilton and P. Grigolini (2001), "The Thermodynamics of Social Process: the Teen Birth Phenomenon," *Fractals* **9**, 193-208.

[139] Scafetta N., and P. Grigolini, (2002), "Scaling detection in time series: diffusion entropy analysis," *Physical Review E* **66**, 036130.

[140] Scafetta N. and B. J. West, (2003), "Solar flare intermittency and the Earth's temperature anomalies," *Physical Review Letters* **90**, 248701.

[141] Scafetta N., S. Picozzi and B. J. West, (2004), "An out-of-equilibrium model of the distributions of wealth," *Quantitative Finance* **4**, 353-364.

[142] Scafetta N. and B. J. West, (2005), "Multiscaling comparative analysis of time series and geophysical phenomena," *Complexity* **10**, 51-56.

[143] Scafetta N., P. Grigolini, T. Imholt, J. A. Roberts and B. J. West, (2004), "Solar turbulence in earth's global and regional temperature anomalies," *Physical Review E* **69**, 026303.

[144] Scafetta N. and B. J. West, (2005), "Multiscaling comparative analysis of time series and geophysical phenomena," *Complexity* **10**, 51-56.

[145] Scafetta N. and B. J. West, (2005), "Estimated solar contribution to the global surface warming using the ACRIM TSI satellite composite," *Geophysical Research Letters* **32**, L18713 doi:10.1029/2005GL023849.

[146] Scafetta N. and B. J. West, (2007), "Phenomenological reconstructions of the solar signature in the NH surface temperature records since 1600," *Journal of Geophysical Research* **112**, D24S03, doi:10.1029/2007JD008437.

[147] Scafetta N. and B. J. West, (2008), "Is climate sensitive to solar variability?," *Physics Today* **3**, 50-51.

[148] Scafetta N., (2008), "Comment on 'Heat capacity, time constant, and sensitivity of Earth's climate system' by S. E. Schwartz," *Journal of Geophysical Research* **113**, D15104, doi:10.1029/2007JD009586.

[149] Scafetta N. and R. Willson, (2009), "ACRIM-gap and Total Solar Irradiance (TSI) trend issue resolved using a surface magnetic flux TSI proxy model," *Geophysical Research Letter* **36**, L05701, doi:10.1029/2008GL036307.

[150] Scafetta N., "Empirical analysis of the solar contribution to global mean air surface temperature change," *Journal of Atmospheric and Solar-Terrestrial Physics* (2009), doi:10.1016/j.jastp.2009.07.007.

[151] Schrödinger E., (1995), *What is Life?*, Cambridge University Press, New York, first published in 1944.

[152] Schwabe A. N., (1844), "Sonnen-Beobachtungen im Jahre 1843," *Astronomische Nachrichten* **21**, 233-.

[153] Shannon C. E., (1948), "A mathematical theory of communication," *The Bell System Journal* **27**, 379-423 and 623-656.

[154] Shaw R., (1981), "Strange attractors, chaotic behavior and information flow," *Z. Naturforsch* **36**, 80-122.

[155] Shidell S., (1964), *Statistics, Science & Sense*, University of Pittsburgh Press, Pittsburgh PA.

[156] Szilard L., (1929), "Ober die Entropieverminderung in einem thermo-dynamischen System bei Eingriffen intelligenter Wesen," *Z. Physik* **53**, 840-856.

[157] Shindell D. T., G. A. Schmidt, M. E. Mann, D. Rind, and A. Waple, (2001), "Solar Forcing of Regional Climate Change During the Maun-der Minimum," *Science* **294**, 2149-2152.

[158] SIDAC, (2003), Solar Influences Data Analysis Center., *http://sidc.oma.be/index.php3*.

[159] Smyth C. P., (1870), "On supra-annual cycles of temperature in the Earth's surface-crust," *Royal Society Proceedings* **18**, 311; *Philosophi-cal Magazine* **40**, 58.

[160] Solanki S. K. and M. Fligge, (1998), "Solar irradiance since 1874 revis-ited," *Geophysical Research Letters* **25**, 341-344.

[161] de Solla Price D.J., (1986), *Little Science, Big Science and Beyond*, Columbia University Press, New York.

[162] Solomon S. et al., (2007), Intergovernmental Panel on Climate Change, *Climate Change 2007: The Physical Science Basis*, Cambridge U. Press, New York. *http://www.ipcc.ch*.

[163] Soon W., S. Baliunas, E.S. Posmentier and P. Okeke (2000), "Varia-tions of solar coronal hole area and terrestrial lower tropospheric air temperature from 1979 to mid-1998: astronomical forcings of change in earth's climate?," *New Astronomy* **4**, 563-579.

[164] Stanley H.E. (1971), *Introduction to Phase Transitions and Critical Phenomena*, Oxford University Press, Oxford and New York.

[165] Stanley H., (1998), "Late Pleistocene human population bottlenecks, volcanic winter, and differentiation of modern humans," *Journal of Human Evolution* **34**, 623-651, doi:10.1006/jhev.1998.0219.

[166] Stevens S. S., (1959), "Cross-Modality Validation of Subjective Scales," *Journal of Experimental Psycology* **57**, 201-209.

[167] Stoneburner R, Low-Beer D., (2004), "Population-level HIV declines and behavioral risk avoidance in Uganda," *Science* **304**, 714-718.

[168] Stothers R. B., (1989), "Volcanic eruption and solar activity," *Journal of Geophysical Research* **94**, 17371-17381.

[169] Strogatz S., (2003), *SYNC*, Hyperion Books, New York.

[170] Sugihara G. and May R. M., (1990), "Nonlinear forecasting as a way of distinguishing chaos from measurement error in time series," *Nature* **344**, 734-741.

[171] Svensmark H. and E. Friis-Christensen, (2007), "Reply to Lockwood and Fröhlich: The persistent role of the Sun in climate forcing," *Danish National Space Center Scientific Report* **3**.

[172] Schwartz S. E., (2008), "Reply to comments by G. Foster *et al.*, R. Knutti *et al.*, and N. Scafetta on 'Heat capacity, time constant, and sensitivity of Earth's climate system'," *Journal of Geophysical Research* **113**, D15105, doi:10.1029/2008JD009872.

[173] Thom R. (1975), *Structural Stability and Morphogenesis*, Benjamin/ Cummings, Reading, MA.

[174] Thompson D. W., (1961), *On Growth and Form* (1915); abridged ed., Cambridge.

[175] Verhulst P. F., (1838), "Notice sur la loi que la population pursuit dans son accroissement," *Correspondance mathématique et physique* **10**, 113-121.

[176] Verschuren D., K. R. Laird, and B. F. Cumming, (2000), "Rainfall and drought in equatorial east Africa during the past 1100 years," *Nature* **410**, 403-414.

[177] Walker G., (1923), Sunspots and pressure, VI. a preliminary study of world weather, *Calcutta, Indian Met. Mem.* **21**, 12 (1915); idem. VIII. Correlations in seasonal variation of weather, idem 24, Pt. 4.

[178] Wang Y.-M., J. L. Lean, and N. R. Jr. Sheeley, (2005), "Modeling the Sun's Magnetic Field and Irradiance since 1713," *The Astrophysical Journal*, v. 625, p. 522-538.

[179] Watts D. J. and Strogatz S. H., (1998), "Collective dynamics of 'small world' networks," *Nature* (London) **393**, 409-10.

[180] Watts D. J., (2003), *Six Degrees*, W. W. Norton & Co., New York.

[181] Weaver W., (1948), "Science and Complexity," *American Scientist* **36**, 536-544.

[182] West B. J., (1990), *Fractal Physiology and Chaos in Medicine*, Studies of Nonlinear Phenomena in Life Science,Vol. 1, World Scientific, New Jersey.

[183] West B. J. and W. Deering, (1995), *The Lure of Modern Science: Fractal Thinking*, Studies of Nonlinear Phenomena in Life Science, Vol. 3, World Scientific, New Jersey.

[184] West B. J., (1999), *Physiology, Promiscuity and Prophecy at the Millennium: A Tale of Tails*, Studies of Nonlinear Phenomena in Life Science, Vol. 7, World Scientific, Singapore.

[185] West B. J. and P. Grigolini, (2008), "Sun-climate complexity linking," *Physical Review Letters* **100**, 088501.

[186] West B. J. and L. Griffin, (2004), *Biodynamics*, John Wiley & Sons, New York.

[187] West B. J., (2006), *Where Medicine Went Wrong*, World Scientific, Singapore.

[188] White W. B., J. Lean, D. R. Cayan, and M. D. Dettinger, (1997), Response of global upper ocean temperature to changing solar irradiance, *J. Geophys. Res.* **102**, 3255-3266.

[189] Wiener N., (1948), *Cybernetics*, MIT Press, Cambridge, MA.

[190] Wiener N., (1948), *Time Series*, MIT Press, Cambridge.

[191] Wigley T. M. L., V. Ramaswamy, J. R. Christy, J. R. Lanzante, C. A. Mears, B. D. Santer, and C. K. Folland, (2005), "Climate Change Science Program 4 Temperature Trends in the Lower Atmosphere," US Climate Change Science Program. *http://www.climatescience.gov*

[192] Willingr W. and V. Paxson, (1998), "Where mathematics meets the Internet," *Notices of the American Mathematical Society* **45**, 961-970.

[193] Willson R. C. and A. V. Mordvinov, (2003), "Secular total solar irradiance trend during solar cycles 21-23," *Geophys. Res. Lett.* **30**, 1199-1202, doi:10.1029/2002GL016038. *http://www.acrim.com*

[194] Wilson K. G. and J. Kogut, (1974), "The renormalization group and the epsilon expansion," *Phys. Rep.* **12**, 75-125.

[195] Zipf, G.K., (1949), *Human Behavior and The Principle of Least Effort*, Addision-Wesley, Cambridge, MA.

Index